Like, Follow, Subscribe

Like, Follow, Subscribe

Influencer Kids and the Cost of a Childhood Online

Fortesa Latifi

G

GALLERY BOOKS

New York Amsterdam/Antwerp London

Toronto Sydney/Melbourne New Delhi

G

Gallery Books
An Imprint of Simon & Schuster, LLC
1230 Avenue of the Americas
New York, NY 10020

First Gallery Books hardcover edition April 2026

GALLERY BOOKS and colophon are registered trademarks of Simon & Schuster, LLC

Certain names and characteristics have been changed.

Interior design by Laura Levatino

Manufactured in the United States of America

10 9 8 7 6 5 4 3 2 1

Library of Congress Control Number: 2025949496

ISBN 978-1-6680-8050-4
ISBN 978-1-6680-8052-8 (ebook)

For E, my dream girl

Contents

Introduction

From May of 1971 to New Year's of 1972, a documentary crew filmed the Loud family of Santa Barbara, California. When the twelve-hour documentary series titled *An American Family* premiered on PBS, the Louds—made up of dad Bill, mom Pat, and five children between the ages of thirteen and twenty—became the first reality television family. Ten million viewers tuned in to watch the Loud family as they lived their everyday, suburbanite lives. They made meals and fought with each other and negotiated the thousands of tiny moments that make up a family's life together. Craig Gilbert, a documentarian and filmmaker, pitched the project to the New York public television outlet where he was working as an exploration of the film style cinema verité, which is meant to capture the intricacies of real life with little involvement by filmmakers. The Louds were chosen, Gilbert said, because they were a middle-class family whose children spanned a wide age range.

In a review from 1973, Anne Roiphe writes that the Louds are "comfortably ordinary" and "sadly familiar. Bill drinks too much and has affairs with other women. Pat usually shrinks from making judgments about anything. And much of the family affluence appears to be contained within a delicate house of credit cards." By the end of the series—during which the camera crew reportedly captured three hundred hours of raw footage that was then edited into the twelve hours that aired—Pat and Bill's twenty-one-year marriage had ended, Bill's business was failing, and Lance, the eldest Loud child, had come out to his parents as homosexual.

An American Family changed television—and media—forever. The blockbuster nature of the hit—and the continued success of the reality TV families that came after the Louds, from the Gosselins to the Duggars to the Kardashians, makes one thing clear: We are fascinated by how other families live. There is nothing more compelling—or more personal—than a family unit. Our curiosity is ignited by the most mundane details of other families: what they eat for dinner, how they divide up chores, the way they speak to each other. The urge to compare our own families to others—and see where we measure up or fall short—feels irrepressible. Are we better than them? Worse? We can't help but watch to find out. Even prosaic moments of another family's life are captivating. We want to know how other people live and when we're shown, we can't look away.

If the Louds are the predecessors of the Duggars and the Kardashians, I also see them as the ancestors of family vloggers and mom influencers, who welcome us into their lives in the same way. We watch as these content creators eat breakfast and put the laundry away and get their kids to sleep. On the screens of our phones and laptops, they get dressed for the day and drop kids off to school and do homework. The kids have tantrums and shave their legs for the first time and decide which college to go to, and we watch. The parents stay together or break up, and we watch. The kids grow, and we watch.

These families open a window into their lives for us to peer through, and through that same opening, the money pours in. Family vlogging and mom influencing are part of the multibillion-dollar influencing industry, with ad money and sponsorships and brand deals stacking up for the families that make their living showing us the details of their days. The most successful family vloggers and mom influencers make millions of dollars a year in a largely unregulated industry that is mostly free from the legal protections under which other, more established forms of entertainment, like child acting or modeling, are governed. In family vlogging and mom influencing, parents have almost total authority—they are not only parents but producers, managers, and cameramen. They decide what is filmed and how often; they monitor social media analytics and negotiate brand deals; they demand that their children participate in reshoots; they sort through messages from fans and haters and pedophiles and stalkers. There is no child advocate on the set of a family vlog. There are only the parents, as the family shifts from its first form into something more resembling a business arrangement.

The children of family vloggers and mom influencers live online. Sometimes their social media presence launches while they are still in utero, as their parents make Instagram profiles for their fetus selves. From the womb, the fetus racks up followers and lands their first brand deal before they're born. Their gender reveal is #sponsored and teased out through several YouTube videos and Instagram posts. Content is spun out of the building of furniture in their nursery and the choosing of swaddles. When the influencer child is born, their birth is announced with an aesthetic photo of their mother, with perfect makeup and loosely curled hair, holding them in the hospital bed. Everything is linked for purchase by viewers, leading to commissions for the parents: diapers and teethers, and later, sippy cups and toddler shoes. The child grows and so does their online footprint: potty training and tantrums and lost teeth and birthdays and first days of school and dance recitals and playdates are all documented. Later, there is content around first menstrual cycles or the cracking of voices as puberty arrives. First dates, kisses, heartbreaks, medical diagnoses, triumphs, and tears are all captured through the eye of the camera and uploaded to social media platforms, where they are spun into views and thus into staggering amounts of money.

These children's privacy is traded for profits; their most intimate moments are broadcast to millions; they lose sight of what is performance and what is not. But they also make more money than many people can dream of—though there are only five states in the US that have laws legally requiring that the money influencer kids earn on these platforms is to be shared only with them. None of those laws addresses the details of working hours or consent. These children are lost in the wild woods of this industry as minors who are at the mercy of both the platforms they serve and the parents who are often holding the camera.

In recent years, the backlash against family vloggers and mom influencers has reached a fever pitch, with parents standing accused of exploitation and manipulation in their pursuit of views and money. Detractors claim that there is no way for children to give meaningful consent to being featured in monetized content and that any parent making money off their kids online is wrong. But I don't think it's that simple. To understand the question of what life is like as an influencer kid, I spoke to the young people themselves, the parents who prop them up, and experts including psychologists, sociologists, and academics. I wanted to hear directly from these kids: What is it like to live your entire life online? What happens when your parent is not only your parent but your boss? And what does it mean to grow up in the shadow of an online footprint that stretches from gestation to toddler tantrums to teenage troubles?

Here is what I found.

1

Dear Diary, Meet the Algorithm:

How Mommy Bloggers Paved the Way

The first mom bloggers didn't know they were changing the world. It was 2005 and the internet was a simpler, less crowded place. Facebook was only a year old (and still only for college students), and Instagram and Pinterest were years away. The conversations moms were having about motherhood took place in women's magazines and advice books and only skated along the surface of the maternal experience. No one talked about postpartum depression or miscarriages or what to do when you started to—just a little, only sometimes—regret having become a mother.

Then, the mommy bloggers entered the scene. They wrote long-form, heart-plundering reflections on pregnancy and motherhood and what their lives looked like after having children. They were honest about topics that had previously only been discussed privately, in hushed tones. They wrote about hating their husbands and struggling with postpartum anxiety and the feeling that their lives were over. It was a revelation. More than that, it was a revolution. It's not hyperbolic to say that mommy bloggers not only changed the way we talk about motherhood but also provided a career path for the influencers of today. When I'm scrolling through Reddit in the middle of the night and reading posts about the guilt that chases moms through every choice they make, I have those mom bloggers to thank. When I read stories about the reality of miscarriages, I know that they were able to be written only because of these original women who broke the silence and began destigmatizing these topics.

One of those women was Catherine Connors. While on maternity leave in 2006, she "fell down a rabbit hole" of mom blogs. (Same.) Within hours of discovering the blogs she would become a fervent subscriber of, she started her own blog called *Her Bad Mother*. In one of my favorite blog posts of hers, she writes about calling her eleven-year-old daughter an asshole, a story that turns into a reflection on raising a girl to be difficult in a world that expects her to be tame. As an academic and new mom, Connors was fascinated by the world of mom blogs. "It was a really radically unprecedented phenomenon of storytelling about motherhood and the family in the public sphere that I figured was going to change everything," she tells me over a Zoom call in which I stay off-camera because I have an almost-four-month-old who refuses to be put down, and my hair is still unbrushed at the time of our 3:00 p.m. conversation. "I wanted to be a part of it, but I was also a brand-new mom with postpartum depression and all the loneliness and all that stuff. I just fell in love with it as a human."

I became a mom in 2024, and it's difficult to comprehend how disparate IRL and internet conversations about motherhood were before the mom bloggers changed the conversation more than two decades ago. Kathryn Jezer-Morton, a journalist who writes a newsletter about family life for *The Cut* and holds a PhD in sociology, explains, "Before there were blogs, people didn't talk about breastfeeding. That was not something that anyone ever talked about—being angry at your kids, just having a vagina, having to get sewn up. Everything that we now really take for granted. It all started with the blog, and it was very revolutionary in terms of what women could talk about." Jezer-Morton says that before mom blogs, conversations about motherhood took place between moms, in women's magazines, and in the pages of advice books. "It was basically like, 'being a mom is hard work!' followed by lots of writing about how being a mom is awesome."

Though mom influencers can now rake in hundreds of thousands or even millions of dollars yearly, the original mom blogs didn't begin as profitable endeavors. "There was no expectation that the blog itself would ever make money," Danielle Wiley, a former mommy blogger who is the founder and CEO of Sway Group, an influencer marketing agency, told me. "It was a creative outlet. It became a community and a place to meet friends at a time when I think a lot of women feel very alone."

When I interviewed women who were part of the original cohort of mom bloggers, this thread ran through each conversation. Mom blogs were a way of holding on to their personhood in the heady, exhausting days of new motherhood. Writing was a way of using their brains for something other than calculating nap times, wake windows, and ounces of breastmilk. It was a way of existing intellectually, outside of the physical drudgery of motherhood. (Yes, motherhood is beautiful. Yes, it is exhausting. Yes, there are days when I feel more like a body that exists to produce milk and rock my daughter to sleep than an actual person.) The blogs were these mom's way of saying: *"We are humans! And mothers! And sometimes it's so exhausting we want to scream! And other times it's so beautiful it feels unbearable! And also, is it normal that sex is painful now?"*

Though I consider the first mom bloggers the predecessors of the current mom influencers, they're related in the same way humans are related to primates. The general makeup is the same and you can see where we're alike, but the differences are stark and clear. The original mom bloggers were writers more than they were what we now identify as content creators. They penned thousands of words on the emotional journeys of potty training or deciding to have a second child. Contrast that with the mom influencers of today, who take beautiful photos and write a few purportedly vulnerable, sponsored sentences to go with it. Rebecca Woolf, an original mom blogger who wrote the blog *Girl's Gone Child* from 2005 to 2017, tells me that mom bloggers started, ironically, as the anti-influencer. "It started as being more of a punk rock scene, fuck the man, we're doing it our own way thing," she says. "The culture has become so . . . beige, for lack of a better word, because you can't monetize it otherwise. You can't sell the fucked up."

But back in the days of the original mom bloggers, because there was no expectation of monetization or worries about coming across as the right amount of aspirational to land brand deals, you could write about the fucked up. That's what Woolf did when she transitioned her regular blog to a blog focused on motherhood when she became unexpectedly pregnant at twenty-three. In writing her blog, she was trying to find a way to both understand the radical change in her life and continue being a writer. And there was a purity and a freedom to those early days that can't be replicated. "We weren't writing necessarily for anyone other than each other," Woolf says. "Our pictures were not good. It wasn't so perfectly curated, and we were super honest and we cursed and we were messy."

On the divergence between mom bloggers and the mom influencers that followed them, Sara Louise Petersen, author of *Momfluenced: Inside the Maddening, Picture-Perfect World of Mommy Influencer Culture*, tells me the earlier bloggers were more writing-centric versus image-centric. They were posting their content on blogs, which are inherently more focused on text than the platforms of today, which are more focused on images. (It's difficult to conceptualize now, but to upload a photo onto a blog, the original mom bloggers had to upload the photo to a different hosting website and then embed the photo into their HTML code.) "It was very much like personal essay writing and plumbing the depths of the maternal experience," Petersen says. "It still felt as though the writing about the kids was in service of exploring the maternal experience."

This is something I've been trying to tease apart: Was it different to be the child of a mom blogger in the early aughts than it is to be the child of a mom influencer or family vlogger now? Was it easier? Was there more privacy? Or is it just another side of the same coin?

Here is what I have come to believe: It is fundamentally different to be the child of a mom blogger than it is to be the child of an influencer. Mom bloggers were focused on their experiences of motherhood to the point that it almost felt like the children involved in the stories were secondary. Mom bloggers weren't telling you about their children; they were telling you about motherhood and sometimes their children were factors in those stories. But mostly, their writing—and the community they created—centered on the women who were raising the children, not the children themselves. Contrast that with the mom influencers of today, where the children are the focal point of their content. "I feel like we were just sharing what our experiences were mothering," Woolf says. And because the first mom blogs started without a path to monetization, the possibility of turning one's family into a profit stream wasn't a motivating force. Can making content based around one's family still be deemed exploitation if there wasn't money involved? And what would it be like to grow up and read a detailed account of your mother's parenthood journey and the role you played in it?

Dylan Wiley, the twenty-two-year-old daughter of former mom blogger Danielle Wiley, can go to her mother's now defunct blog to read details about her childhood. When I get Dylan on the phone, I ask her what she remembers about those days. Dylan tells me that the clearest memory she has of being part of her mother's blog was when she was around eight years old. Her mother was recording a video tutorial on how to make a roasted chicken and had asked Dylan to help her. "She had a little Post-it note on her head that said what I was supposed to say," Dylan says. "She wanted me to help her make the chicken and told me what I would have to do for the video." The image Dylan paints is striking—and there's a temptation to view it sinisterly: the mom blogger roping the young daughter into the creation of content and scripting her lines. But is that how Dylan sees it? "I loved cooking with her," she says simply. "I just always knew [the blog] was a big part of her life." I search through Danielle's blog to find the video of Dylan helping her mom make chicken. In the video, she has gapped teeth and bangs swept across her forehead. "Hi! It's the Wiley family," she says in her little kid voice. "And we wanted to teach you how to make roasted chicken." As Dylan details the steps in the recipe, she stumbles across her words and giggles. It's endearing and sweet, a moment of her childhood frozen in time.

Now, as an adult herself, Dylan appreciates the detailed record of her childhood that exists online. "I love reading back . . . and just reminiscing," she says. "[My mom] would talk about parenting from her perspective. I always found that really interesting, because it seems like moms are professionals at what they do. So knowing that this was also her first rodeo and just getting a glimpse into what motherhood was like through her eyes has always been really interesting for me." Amidst her mother's blog archive, Dylan particularly loves the posts that detail her first steps, her first day of kindergarten, and how hard it was for her mother the first time she went to sleepaway camp. "It just allows me to get to know her on a level that I would not have been able to before," she says. "Getting to read things through her eyes . . . firsthand has been really lovely."

It does sound lovely. For a moment, I think about what it would be like to be able to read about my own mother's musings on parenthood. There's beauty to having such a thorough record, and part of me is almost jealous of Dylan. I ask her if there was ever anything her mother shared that made her uncomfortable. No, Dylan tells me.

Her mother didn't write about potty training or tantrums or anything that strikes her as too intimate. And because of the format of the blog, which was largely text-based instead of relying on videos and photos, her experience as the child of a mom blogger was fundamentally different from that of the child of an influencer. No one recognized her in public or had access to thousands of videos of her daily life. The blog was just a creative outlet for her mother (albeit one that transformed her mother's career as she moved from mom blogging to representing mom bloggers in her work as the CEO of her own marketing agency).

What a wonderful bow on the story of mom bloggers Dylan's experience would be. But that wouldn't be the full story.

When I call Shannon Bird, a Mormon mommy blogger based in Salt Lake City who has five kids between the ages of five and fourteen, she laughs brightly into the phone and tells me she just finished an appointment at her plastic surgeon's office. She requires almost zero prodding before launching into the story of her own ascent into mommy blogging. "I am a freaking OG mommy blogger," she says. "I feel like the new girls on TikTok, they've done it for like a year and they're famous. And I'm like, no guys, we were going back to basically the newspaper. We had to start writing, and taking photos, and uploading and resizing photos. I'm from the dinosaur age." (Not quite—she started blogging in 2011—but I get her point. Compared to the internet of today, the internet of 2011 does sometimes feel prehistoric.) She remembers traveling for sponsored family vacations and bringing filming and camera equipment with her. "Now it's like, everything is on your phone. Honestly, with the new girls coming in, they get famous quicker. Do you know what I'm saying?" I assure her I do. "It was like grunt work," she continues. "I [was] competing against

Google, competing against *The Wall Street Journal*. We were basically journalists. We were doing diaries online."

Shannon talks quickly and animatedly. What set her apart, she tells me, was her honesty. "I documented a miscarriage I had," she says. "You should have seen the claws come out. [It was like] 'You do not talk about this stuff online. This is so inappropriate. This is between you and your doctor. This is a personal matter, you shouldn't talk about this.' Why not? I questioned why. Why can't we?" Shannon lived her life online; loyal readers waited for her latest dispatches, and brands she was working with waited for her to post sponsored content, so disclosing her miscarriage was more than just personal—it was business. "And I think it's fair to the audience, like, hey I announced I'm having a baby. That baby's no longer coming. I think women should be able to talk about this. And I had brand deals and they needed to know why I was not going to be able to post pictures for a few weeks because I was mentally and physically not in a good spot. [But back then] it was taboo to talk about it."

I want to read Shannon's blog post about her miscarriage, but I can't because it is no longer online. In fact, her blog doesn't exist at all anymore. She took it down a few years ago when her eight-year-old son came home from school crying. When Shannon asked him why he was crying, he wouldn't answer. Eventually, he asked how much money she makes from her blog. "I was like, 'Well the blog is dying; the written word isn't being read,'" she said. "He's like, 'Would you consider taking it down?'" In computer class, her son explained, his peers found Shannon's blog and used it to make fun of him. They read aloud from his birth story and printed out photos of him as a toddler. Emily Kline, PhD, an adolescent psychologist, said this is one of the concerns she has for kids of influencer parents. "On a very concrete level, their classmates may find those accounts and watch them and say about them like, 'Oh, I watched the video from five years ago where your mom said that you pooped your pants at the playground, LOL.' That's just not something that a third grader would want immortalized on the internet."

That this was happening to her son, whom Shannon describes to me as a "cool, popular kid" who is a quarterback on the football team, shook her. She says she immediately took the blog down and went private on Instagram (though at the time I'm writing this, her profile is public, she has close to 90,000 followers, and her most recent post is an impeccably posed photo of her family.)

Talking to Shannon about her journey online is a study in contradictions—she seems sincere when she tells me about deleting her blog because her son was being bullied but then talks about how difficult it still is to rope her kids into sponsored content. "My oldest son [who is thirteen years old] starts it, like saying 'How much are we getting paid? My friends are going to see this, block all my friends,'" she says. "I ignore it. And then my daughter [who is eleven years old] will throw things to the ground and be like, 'I'm not wearing it, this is dumb,' and she'll walk off. And then my husband will go grab their hands and be like 'Guys, just do it,' and then it's always bribery."

What does she bribe them with? Disney World, she says. And what does she do when they don't want to participate? "I'm like, 'You guys can do this for me. I literally spend sixty hours a week driving you to sports, you guys can do one photoshoot. No mom is doing what I'm doing for you guys.' I'm like, 'Do you know how much your sports cost me every year? No, neither do I. All I know is it's a lot.'" That's not to say that Shannon doesn't try to limit the participation of her children on her Instagram page (though it seems like those limits stem from how frustrating it is to get them to play along rather than any concerns about privacy.) But what is she supposed to do when she's paid eye-watering amounts of money for sponsored content? Recently, she was offered $12,500 for a grid Instagram post of melatonin gummies for kids. She considered not doing it out of concern over the possible backlash from commenters who may suggest she's drugging her kids but, come on—it was $12,500. Which just so happened to be the cost of the boob job she was planning on getting done. "I did an ad on it. Of course, all the mean comments came in, but I'm like, *Free boobs, free boobs. This is for free boobs.*"

And she understands why her kids don't want to be involved—"They've been doing it since the minute they got out of the womb," she says, "they just dread it"—but what other career would offer a stay-at-home mom this much money? She remembers a post she made about diapers that paid her $4,800 monthly on top of never having to pay for diapers again. And look at the beautiful life she's given her family, free of financial stress. She can't help but remind them of that. "This is how we're paying for your college. You'll be so happy when you don't have to pay for college. You'll be so happy when you get a Range Rover for your sixteenth birthday," she says. "Hello?! Do you think those things just happen?"

At the height of her mommy blogger fame in 2019, when Shannon was making $16,000 a month, the family was constantly traveling. The children's school would call her and tell her they were missing too much class, but she had brand deals expecting her in Mexico and the Dominican Republic and New York. And isn't travel valuable too? she reasoned. It didn't seem to be to her kids. "We were flying all the time. My kids, they just want stability. At the end of the day, they don't care about New York. New York will always be there, but sixth grade won't. So what they actually want is a childhood. They want the *Leave It to Beaver* lifestyle," she says and the frustration is evident in her voice. "Is that what you want?" I ask. "I've accepted it," she says with a sigh.

As Shannon grapples with her decision to involve her children in her career in such a central way, she just hopes they understand where she is coming from. She imagines what her children will say on a therapy couch someday. Will they remember her as a mother who only cared about money and brand deals? Or will they try to look at her decisions through an empathetic lens? "You guys always had Disney World at your fingertips. You guys were getting paid to go to amusement parks and to go to Punta Cana," she says to me, imagining a conversation with her kids I wish I could listen to in real time. "Was I overexposing you? Maybe. Were you guys also living a life I never even had? Yes." In the end, she hopes her children give her the same grace she tries to give her parents—that they did the best they could with the tools they had at the time.

Shannon's blog has been taken offline, and the only platform she currently makes money from is Instagram. Her revenue comes through the Instagram bonuses program that pays some users for the amount of interaction they get on their posts. She also gets commission from any products she pushes (her words, not mine), and she trades social media content for products. Shannon recently got a $15,000 nose job for free in exchange for posting about it. She gets $5,000 worth of injections per year in a similar exchange. Sometimes she is sent on sponsored trips. "It's not what it was, but it's still good enough to keep me around," she says. At her most successful, she made $19,000 a month. Now, she nets between $3,000 and $5,000 a month. The change in income has been so drastic that her husband has had to go back to work after quitting his full-time job during the heyday of Shannon's blogging.

As our conversation winds down, she stops for a moment and asks if she can ask me a question. I tell her she can. "Do I seem confused?" she asks.

Listening back to the transcript of our conversation, I can hear the silence between us as I think. "I think you seem pulled in several different directions," I say carefully.

"Yes," she sighs. "That's an educated way to put it."

Natalie Jean Lovin's blog was originally called *Nat the Fat Rat.* Her first post was about a trip to Target on Atlantic Avenue in Brooklyn ("I loaded up my arms. The plastic handles cut into my skin. *Now is not a time for pain*, I told myself. *Now is a time for heroes*"). It was 2005 and Natalie had moved to New York City for her husband to complete a job training program.

That first year, she wrote 24 posts and by 2010, when her son Huck was born, Natalie was writing 248 posts a year. She wrote about the years of infertility she endured before having her son, the retail jobs she found unfulfilling, the struggles of moving across the country multiple times, the joy of her pregnancy, and the wonder and difficulty of motherhood.

In a piece for *The Cut*, Lydia Kiesling traces the trajectory of Lovin's blog:

> Over time the tenor changes—the look of the blog becomes sleeker. Nat the Fat Rat becomes Hey Natalie Jean. She gets sponsorships, a beautiful loft in Brooklyn with exposed brick. She writes a book. Once a quiet defender of her faith, she leaves the LDS church (her husband does not). She struggles with secondary infertility. She gets a tattoo. And then suddenly they are back in Moscow, [Idaho], where Brandon gets a professorship at the law school, and Natalie gets chickens. In April 2016, she and Brandon get divorced. She shutters her blog, the archives too.

The blog and the archives are back now, but Natalie isn't, not really. She's on Instagram, where she occasionally posts to her nearly 40,000 Instagram followers. Her bio reads, "Mother of chickens. First of her name. Kind of a dork. Used to be cool. Looking for an agent—DM if interested" with a link to her website. I message Natalie for months before she agrees to talk to me. Maybe it's because I've spent hours combing through her log of the most personal moments of her life—infertility, birth, and marriage troubles—but I feel like I know her. Ever since leaving the blogging world, she says, she's tried to "studiously avoid" it. Why? I ask. "Well because it's a dark place to be, watching other women do things," she says. "Obviously, we're all performative. We're all putting out there either what's great or what's awful but funny. And that is a highlight reel that no one lives every day, and you can get sucked into it and you can find it really inspiring." Now, she says, that purported perfection disturbs her. "It just puts me back there, where I am only as beautiful as I look in a photo and I'm only as worthy as what I can pull from. I know how it feels to look shiny and to look great while inside being so desperate for something. But ultimately, I spent so much time living in there that even a wish of it makes my skin crawl."

Now, Natalie occasionally posts on Instagram, but it's nothing like it was when her blog was at its most popular. Some of her contemporaries are still massively popular, like Jo Goddard, aka @cupofjo (302,000 Instagram followers), who runs a lifestyle website. Natalie, on the other hand, has let go of blogging, gotten a day job, and mostly exists off the internet. But sometimes she misses the audience, her viewers.

"There is a part of me that regrets that I let that go," she says. "That part of me that's always receiving attention for myself and what I looked like and how I sounded and how people were perceiving me."

What's hard about being on the other side of blogging? I ask. "It's just not knowing. It's just not having tangible proof," she says. "It's a lack of data." Lack of data about what?

"About—I don't know, whether or not what I'm wearing is cute, or if my makeup is serving, or if the joke I thought was funny is actually funny. Without having that immediate feedback about [whether] what I've done is acceptable, I'm just wandering through this abyss by myself being like, 'Okay, I bought these dishes. Do I love them? Maybe.' But if someone else were to say, 'These are perfect for you,' I'd be like, 'Yes, I do love them.'"

If the original mom blogs were mostly authentic community-oriented spaces, what changed? How did we get from Catherine to Shannon? It began with the banner ads that mom bloggers were able to feature on their blogs in exchange for payment. The decision to monetize their blogs wasn't an easy one; there was a sense, from the readers and sometimes from other bloggers themselves, that the act of monetization would ruin the purity of the enterprise. Looking back, maybe it did. But what was the alternative? That these women were not paid in any way for the considerable labor they were performing? It seems almost too perfect that there is a part of us that wishes this is the road that was taken. Asking women for labor without payment? Groundbreaking.

Danielle Wiley tells me about the moment banner ads were announced at the 2006 BlogHer conference. BlogHer is a media company founded in 2005 that hosts regular events for mom bloggers, and that year, the conference was at a hotel in San Jose, where attendees were sitting outside by the pool, sprawled across lounge chairs and at tables. They were awaiting a session when a BlogHer representative came outside and announced that any blogs that were part of the BlogHer universe would be eligible to run banner ads.

"I remember it was dark out, and I remember people screaming at each other," Wiley says. "It was blogger versus blogger. It was super antagonistic. We started this to have a creative space and you're going to monetize it and you're going to ruin it and you're going to change it."

At the time, Danielle remembers thinking the detractors were "crazy." Shouldn't the women who were pouring so much time and energy and creative effort into these enterprises be rewarded for that? She still believes they should have been. But there's no denying that the switch to monetized content changed everything.

In her book, *Extremely Online: The Untold Story of Fame, Influence, and Power on the Internet*, Taylor Lorenz details the mind-bending journey from small mom bloggers to a multibillion-dollar industry. "It was a fraught decision to try to make money online," Lorenz writes. "Mommy bloggers' advantage was their authenticity. Ads were seen as tacky and inauthentic, a sign of selling out. Even though nearly every top mommy blogger worked on their blog full-time, they and their audiences appeared to internalize negative stereotypes about the economic value of the work they were doing."

The backlash didn't stop the train from leaving the station. Banner ads were introduced, and soon after came the sponsored content (or, in today's parlance, #sponcon). As Lorenz continues in *Extremely Online*: "Ultimately, mommy bloggers, more than any other group, created the model that content creators and platforms cultivated in the decades that followed. They were among the first people to commodify themselves online, to post candidly about their personal lives—and then monetize." But how did we get from that cottage industry of mom bloggers to the mom influencers and family vloggers of today, some of whom pull in millions of dollars per year? And when did the focus switch from the experience of motherhood to the lives of the children?

The ideal mom influencer went from a woman who was an accomplished writer and could pen long essays about the meaning of motherhood and the drudgery of day after day of dirty diapers to a woman who could make her life look perfect and beautiful and tell us maybe a little too much about her children along the way.

I think of it this way: When I ask people who their favorite original mom bloggers were back in the day, they name the mothers. They say they loved Dooce because of how frank she was about postpartum depression, or they loved Natalie Jean for detailing the devastation of a miscarriage. But when I ask people now who their favorite mom influencers are, they tell me they love Savannah LaBrant because her daughter Everleigh is so cute, or they favor the kids from the Family Fun Pack. Even when the platforms belong to the mothers, the focus is the children. I don't know that we've fully reckoned with what that means.

Mommy bloggers changed the internet forever. They made it possible to make a living (and sometimes, a generous living) off sharing your personal life and monetizing it. But they also opened a door that can't be shut.

2

A Filtered Life:

Fame, Family, and Clickbait

"**Let's make a period kit** for my daughter," says a grinning @whataboutaub (1 million Instagram followers) in the Instagram video. "My oldest just turned twelve, which means the big moment is just around the corner. Her first period." The video is upbeat and cheerful, marked with the signature #ad in the caption. As she speaks straight to the camera, Aubree explains that she's partnered with U by Kotex, and she's glad to be handling periods way differently than they were handled back in her day, when it was "hush-hush and awkward." While she talks, Aubree packs perfume and tissues into a basket sitting on her lap. "I remember how much I hated wearing huge pads that felt like diapers, but with these U by Kotex Balance Teen pads, my daughter won't have to deal with that because U by Kotex Balance sized for teens are ultra-thin, shorter, and narrower than their Balance Core line version. How great is it that we get to do this differently?" We get to talk about periods openly, Aubree says. (So openly that it can be made into sponsored content.)

I wish I could tell you that Aubree Jones's sponsored post about her daughter's imminent first period was the only one I've seen of that kind, but it's not. In the world of mom influencers and family vloggers, anything can be made into sponsored content—first menstrual cycles, medical diagnoses, potty-training routines. Nothing is too personal to be #sponsored and everything is potential content.

That's something Iris (a pseudonym) saw up close when she was a nanny for a prominent influencer family from 2016 to 2020. (Though in our phone conversation, she's quick to tell me that she was only their weekend nanny, not their full-time caregiver.) As a nanny in a major metropolitan area, Iris was used to working for wealthy people, but working for an influencer family was much different. "Included in my nanny duties was helping with the kids [so the parents could make] content, which honestly just made me so incredibly uncomfortable. Having to help them try to get the kids to do a shoot for vitamins or something seems so trivial when you're trying to calm down a two- or three-year-old," she says. "Especially having to calm down a tantrum in a kid who clearly needs a nap and a snack and whatever."

Something that surprised Iris was how quickly the children became accustomed to the creation of content. The eldest child was especially cognizant of cameras. "When he was a toddler and he was going to preschool, he would say to me, 'Are you recording?' Or he would be like, 'Wait, I'm going to do this. Can you take a picture for my mom?,' and I'm like, 'Of course, sure.' And then he would have me retake it. He's like, 'No, that isn't good. I want to do another one.' It was just so disconcerting that I saw how quickly a kid who barely has grasped potty training is so aware of my phone, of a camera, how to smile, how to pose." Once, the child told Iris he had to have three birthday parties because his mom had to film two of them. "One was a sponsored-birthday type deal and then the other was for filming content and then the last one was finally a normal birthday that you'd have for a toddler." Other times, the child talked to Iris about PR and unboxing videos "which was weird, talking about that to a little kid. It was very bizarre."

While Iris nannied for the influencer family, she became a part of the influencer world. She met nannies of other family bloggers and mom influencers who helped run the households behind the scenes. Online, the nannies were not part of the picture the moms were selling. Iris listened to the creators speak on podcasts and Instagram Lives; fans and hosts asked them how they balanced it all, and the women answered that they relied on effective time management and caffeine and intense scheduling. What they didn't reveal was that they had nannies or any kind of childcare. Instead, they made it seem like their picture-perfect lives were made that functional and that aesthetically pleasing by their own efforts alone.

As a woman of color who worked with childcare providers who were largely also women of color, it made Iris uncomfortable to see their labor erased. "I was like, 'What in the white feminism?' I really hated that I knew there was this really exploited class of women running throughout the city while these moms are sitting here making it look easy and showing off these very glamorous and wealthy lifestyles, when I knew that their nanny was underpaid." Aside from the labor issues, it bothered Iris on another level too. She would hear regular moms talk about mom influencers and family vloggers and how perfect their lives were, how they could handle everything in their houses and careers on their own while looking good. And it made those women feel like shit because if the mom influencers could handle it and make it all look beautiful, why couldn't these women?

But what they didn't know, and what Iris knew all too well, was that the influencers had nannies and housekeepers around the clock. These average moms were comparing themselves to a standard that didn't exist. (Even as I write this book and interview people like Iris, I find myself doing the same thing. Why can a mom influencer wrangle her nine children and keep her hair perfectly waved and wear a beautiful dress while baking sourdough from scratch while I'm sitting here in a milk-soaked T-shirt, not even sure if I brushed my teeth this morning? And I know the image these influencers curate of their motherhood is not real! But there's still part of me that believes it is—and that I'm falling short.)

This is something I've heard over and over from sources—that the mom influencers and family vloggers who make content around doing it all themselves actually have a ton of hired help. Consider what Lisa, the neighbor of a prominent vlogging family told me: "They outsource all of this stuff—they have a cleaner and nannies on regular rotation. But they make it seem like, 'Oh well, if you just plan your Sunday evening and have your meeting with your partner for the week, then everything's going to go smoothly.' But that's not how it works for most people." It strikes Lisa as particularly dishonest because the vlogging family she lives next to sells courses based on how to organize your life and your household as a parent of multiple children. What's not included in those courses? Any mention of their nannies or cleaners.

Let me be clear: I don't think there is any shame in hiring people to help you run your household. What I do find troubling is hiding the people helping you while selling a version of your life where you are doing it all yourself. The promise—that if viewers just follow your lead, consume your content, and buy your courses, they can have as perfect of a life as you appear to—is based on a lie. And that's not the only lie. One mom influencer who requested to remain anonymous told me that she knows creators who rent hotel rooms to film in while passing them off as their actual houses. Another source, who worked as a doula for a vlogging family, said their content is based around living on a sprawling, hundred-acre farm. In reality, they live on a couple of acres. "You would not know that from watching their page," she says.

The influencer Iris worked for never shared in her content that she had nannies 24/7. (Remember, Iris was only the weekend nanny, with another nanny working during the week and one on call for any other time.) Once, the mom Iris nannied for left her eldest son with the main nanny for the weekend and tasked the nanny with potty training him. The nanny did an intensive few days of potty training, telling Iris all about it. Later, the influencer mom posted sponsored content about potty training, with tips and tricks for other moms even though Iris knew she had nothing to do with potty training her son. It didn't matter—she could still monetize it.

Watching how the dynamics of the family shifted during filming troubled Iris. She was specifically worried by what she called the father's "short fuse." "Literally just them doing kid things" would be enough to irritate Dad, Iris said. If the toddler got his clothes dirty or whined during a filming day or acted, you know, like a toddler, the dad would be in the background rolling his eyes. Though it made Iris uncomfortable, her job as the nanny was to support the parents in their parenting choices—even if that meant trying to get little kids to comply with filming content. "You may not be agreeing with the parents' choices, but that's not your job," Iris said. She started to figure out ways to get the kids to be more willing to work with her. "I would say, 'We're going to help Mommy with this, then after that, we're going to the playground.'" In a way, it was like any other nannying job—by manufacturing her own excitement, Iris could hope to keep the kids on track. If a photoshoot was fall themed, she would hype the kids up by telling them they would get to play with pumpkins. If the content involved bubbles, she would tell them ahead of time to try to build a sense of excitement and anticipation.

To the mother's credit, Iris says she always tried to get the filming done as quickly as possible. She also tried to make it fun, playing with the kids while creating the content that would later become monetized. But Iris began to notice a difference in how the parents interacted while the cameras were on and when they were off—and so did the eldest child. "With kids, they have a specific way they want to play," Iris said.

"He could not quite understand why sometimes he could play however he wants to play [and then other times] he has to play in a specific way with his mom [for filming purposes]." While Iris is telling me this, I'm thinking about how difficult it would be to get a toddler to play with a specific toy in a specific manner at a specific time. What would I have to use to bribe a kid to do that? And how irritated would they be?

Sometimes the eldest son would be tasked with playing with a sponsored toy, like a dump truck, but what he really wanted to play with was his blocks. Trying to get a toddler to understand why he had to play with his dump truck—and achieve the perfect aesthetic while doing so—was difficult. Dad would get upset and frustrated, Iris says, while Mom would try to play the peacemaker and negotiator. "She'd be really fast to step in and try to be bribing, like 'Okay, well after this, you can have X, Y, and Z.' [But it was] in this overexaggerated way that she's trying to interact and play with the kid, which would weird him out because kids know. They realize this is not usually how Mom is. It was just weird and uncomfortable for him."

In the research paper "Famous at Five: Risk Assessing Digital Child Labour," Dr. Francis Rees, a lecturer in law at the University of Essex, writes of moments like the ones Iris described: "These interactions require a great deal of skill and balance from the parent, ensuring that the child complies with the instructions, that the task is completed satisfactorily and to time; all while according with directions that it looks 'natural' and 'relaxed' and that the child is 'really having fun.'"

Clinical psychologist Emily Kline, PhD, says there's a strange disconnect for the kids of influencers, where they don't have the critical thinking capacity to differentiate between a private self and a public self. If a child has to pretend to do one thing for the camera when they know they would do something else if the camera weren't there, they can be confused. "I think that requires a lot of abstract thinking," Kline says. "Obviously, children don't think abstractly. They're like, 'Are we going to fake it for the cameras so that Mom has content to post?' How do [they] make any sense of that?"

In my conversation with Kline, I tell her that I'm often struck by how strange it is for an influencer child to have few to no boundaries between their home life and their work life. If their work is their life and their home is the set, how can they institute a meaningful boundary? Kline says, "Children's work is development. It's learning to use the toilet and make friends and separate from their parents for eight hours of the school day and tie their shoes and resolve conflict and help out. That is the work of childhood." Put more simply: Influencer children shouldn't have to figure out the boundary between their work lives and their home lives, because they shouldn't have work lives. But they do—and their schedules can be grueling.

Brandon Stewart, the founder of Brandon Studios and a former producer at AwesomenessTV, which works with kid influencers, laughed when I told him that most family vloggers I spoke to insist that they film only a few minutes a day. I asked him why he was laughing.

"It's like running a network," Stewart says. "You have to develop, and you have to green light it, talk to the talent, which is their child, convince them to do it. You have to film it, and then you have to get it edited and then put it out. Even if the camera may be only on for this amount of time or that amount of time, they have to promote it on social media. They have to be present with the fans. That is a whole job in itself. No, I don't want to call anybody a liar, but I've been in the industry, especially this kidfluencing industry, for a long time." To put it even simpler: Anthony Ambriz, a YouTube strategist from Utah who has worked with many of the most successful family vlogging channels, says that when it comes to big channels, there are content requirements every day. "It's daily. Videos are filmed all day. They're shot to be released the very next morning."

The entire experience of nannying for the influencer family soured Iris on influencers in general and left her more cynical about what she sees online. Even now, years after she left that family, she wonders: *Will those kids be some of the ones who eventually tell their stories of what it was like to grow up with influencer parents? Will they resent how much of their life was documented online?* She still thinks about them. She hopes they're okay. And they look like they are, at least where she can check on them, framed perfectly in the squares of their mother's Instagram page.

Rachel is a teenager whose mother is a Facebook influencer with hundreds of thousands of followers. Her mother started posting on Facebook in 2010 but gained popularity around 2021, when she started regularly posting family vlogs and hosting near-constant live streams. Sometimes, her posts are faux-inspirational artless memes that Boomers are particularly drawn to ("Grow where you are planted," etc.), but it's during her live streams that Rachel's mother really shines. Her viewers invest "stars" in the streams and ask for shoutouts, and Rachel's mother, ever benevolent, grants some of them. Sometimes there are prizes to be wished for or won.

Rachel tells me her mother films everything: the family's meals, whenever they take a walk, any time they throw a party. If something funny or interesting happens when she's not filming, she takes her phone out and requests a reenactment. The constant filming is "disruptive," Rachel tells me, and has left her with a simmering fear that anything she did would be recorded and posted online. It also changed the nature of her relationship with her mother. Rachel keeps birds as pets and describes what happened when one of her birds passed away. "We were having a funeral and burying it, and I was crying, and all she did was shove a camera in everyone's face and wave in front of the camera in a chipper voice, saying, 'Bye, bye!'" Rachel remembers. Another time, her mother also filmed an entire tour the family took on Alcatraz Island, which Rachel found distasteful.

Rachel has confronted her mother about not wanting to be part of her content, but her mother's reaction left the issue unresolved. "If I tried talking back to say, 'Don't film' or 'Hey, stop,' she'd start crying, telling me it's her job and it's how she can support me," Rachel says. "It's not. She has a full-time job." As a Facebook influencer, Rachel's mother makes markedly less money than other influencers do—even with hundreds of thousands of followers, Rachel estimates that her mother makes less than $100 a month. When she tells me this, I'm momentarily at a loss for words. All of this work and interfamilial strife and obsession with content creation for less than $100 a month? Yes, Rachel says. Before Facebook changed its monetization guidelines, her mother made up to $400 a month. But it was never life-changing money for their family.

So if it's not money, what does Rachel think her mother's motivation is? She's not totally sure. On one hand, her mom has always been a person drawn to side hustles like getting involved in multilevel marketing schemes or reselling things on eBay. On the other hand, this feels different—like she's getting something from her Facebook platform that might be more rewarding than money. "I guess it's a hobby she's really invested in," Rachel says, though she sounds uncertain, and I feel almost guilty for asking this question because what I'm really asking is: What does your mom get from Facebook that makes this family tension worth it?

There's also the fact that Rachel's mother is an immigrant and her Facebook platform is a way for her to connect with family and friends she left behind in her home country, as well as other members of the diaspora in the United States. "I think, really, that she misses her home," Rachel says, and it almost breaks my heart how tender her voice sounds. "It's a way for her to connect with people from her home country. That's who most of her fan base is."

Rachel says her mother has always been someone who values the outward appearance of her family. Facebook just gives her a way to platform that perfection—while also intensifying the desire for it. "Being online really amplified her need to look more perfect than we already were," she says. "Her values are very much how we look on the outside."

The constant filming is exhausting to Rachel, who is always "paranoid" that there's a camera trained on her. "With my mom, who films everything, who live-streams everything . . . there's a relief when she's not around," Rachel says. "I haven't heard a 'How's your day?' in a while. She just calls me over to show me her followers on TikTok. I didn't think I could get jealous over people on the internet, but when she streams, I do. She'd rather trade stars with strangers than spend time with her daughter."

Rachel's mother's obsession with content creation changes not only how Rachel experiences life but the actual kind of experiences she has. When she turned sixteen, she wanted to have a Sweet Sixteen birthday party but decided not to because she knew her mother would insist on filming the entire thing. "The only way for that not to happen is to just not have one," she says. "I stopped celebrating birthdays." Once, her mother posted a video of Rachel's brother sleeping; his friends found the video and teased him endlessly, which added yet another worry to Rachel's list. "I feel paranoid at times. I did not want to sleep around her." The constant filming also colored the quality of her own memories. "Sometimes all I remember from a certain trip is my mom recording instead of paying attention to our family. I feel like she missed out on some parts of me, and that's her fault because I've tried instigating to spend time with her, and she would rather sit back and film videos. Sometimes I'd try starting a conversation and she'd be editing and have me wait till she's done. Sometimes I go over to my friend's house and I notice the difference of like, 'Oh hey, her mom isn't on her phone at all compared to mine.'"

Toward the end of our conversation, Rachel tells me her mother's social media presence leaves Rachel embarrassed. It's uncomfortable to see her mom shoving a camera in people's faces or insisting on spending hours live-streaming and recording and uploading. It's a violation, she says. I tell her I'm sorry.

"It's all good," Rachel says. "It's life. Well, it's my life."

When I first stumble on @jamienotis, who has 1 million followers on Instagram, I watch a story of her, her husband, and her twin newborn sons waiting to be seen at the hospital. "Huxley is so fussy" reads the text over a video of mom Jamie rocking and shushing the crying baby. "I know, buddy," she says as she soothes baby Huxley (who is still so young that he has that scrawny look newborns have before they regain their birth weight) and her husband films. The hospital, Sarasota Medical Center, is geotagged. The next story is Jamie and the doctor arranging Huxley for an ultrasound, tucking pillows around his tiny body. In the third story, the doctor proclaims that Huxley is fine as a tech wipes what looks like ultrasound goo off his hip. Huxley is drinking from a bottle and naked from the waist down, and Jamie's husband zooms in on his face.

The other content on Jamie's profile maintains the same level of extreme intimacy and disclosure. One of her pinned videos opens with Jamie holding her older son on her lap. He looks about four years old and very, very sick. He's wearing sweatpants and no top. His cheeks are flushed and his lips are parted like he doesn't have the energy to close them. His eyes are unfocused and dazed, lids half covering them.

"Hey," Jamie says. "Hey. Hey, baby. Are you okay?"

Her son doesn't respond and it doesn't seem as though he's heard her, or if he has, that he's capable of responding.

"Can you get me a wipe for his mouth, Doug?" Jamie asks of her husband. Doug, holding the camera, says "yeah" and grabs one. "Something's off with our son," Jamie says. "Yeah, I know," Doug says, still holding the camera.

"Hendrix," Jamie says, running her hand through his hair. "We're getting in the truck and going," Jamie says to her husband, who is still filming. "Find the hospital."

"It's twenty-seven miles away," Doug says as Jamie wipes Hendrix's mouth.

"Twenty-seven miles?" Jamie says and you can hear the panic rising. "Are you kidding me right now? Something's off with our son." Her voice is cracking, and I feel sick to my stomach. I shouldn't be watching this. No one should. Why is it pinned to her profile?

"Hendrix," Doug says, grabbing his son's arm in what looks like an attempt to get his attention. Hendrix is bleary-eyed and unfocused. He hasn't moved during the entire video. "Something's off with our son," Jamie repeats. "Something's off with our son. Doug!"

"Yeah. Yeah. Okay, hold on," Doug says, and he is *still filming*. In the background, another child starts crying. "It's okay, Henley," Jamie says. "Hendrix," Doug says again. "Hendrix," Jamie says, and the child in the background continues crying. "He's going to seize, I think. Call 911. Please, Doug. Call 911."

"Where's your phone?" Doug asks, and he steps back, the angle widening to include the crying Henley. Hendrix still hasn't moved. "Call 911!" Jamie repeats, and the video—finally, mercifully—ends.

Not much surprises me anymore after years of reporting on family influencers. I've become inured to a lot of content that would shock other people. But this video gets to me. It leaves me shaken. What possessed Doug to keep filming? Why didn't he stop the video and call

911 using his own phone instead of looking for Jamie's? Why didn't he comfort his other child, who was crying and watching her brother look lifeless? Why didn't Jamie tell him to put the camera down? And why did she post it?

Maybe the caption has some answers? "This was just moments before my baby became unresponsive, stopped breathing and his lips turned blue," Jamie wrote. She details the day—that the family had gone for a hike in the Badlands of South Dakota and Hendrix had seemed completely fine, though he had wanted to be held "which was off for him." Jamie says her "mama instincts" were telling her something was off. Later, after dinner, Hendrix had a slight fever, so they gave him Tylenol to try to ward off a febrile seizure, which he'd experienced before. "With his previous seizures he'd twitch a lot, moan lightly, & foam at the mouth a little or throw up a little. This time, his legs twitched briefly, his eyes had no focus and rolled to the back of his head, and then he stopped breathing for what seemed like forever. I tried to stay calm for my daughter, but when I looked down at my baby with a grey face, blue lips, limp body— completely unresponsive, not even twitching—lost it. I genuinely thought he was dying. I ran out of our RV and screamed 'help!!!!' like a maniac bc I thought we'd need to do CPR. Doug was on the phone with 911 the whole time & they worked so hard to get to us quickly," she wrote. (Was Doug on the phone with 911 the whole time? He didn't seem to be while he was recording the video Jamie later posted.)

She finishes her lengthy caption with "Once Hendrix became responsive and stable again Doug looked at me and said, 'you have some instincts!' Mamas, even if you think you're being 'paranoid' or 'anxious'—trust your instincts."

The urge to share detailed accounts of medical crises isn't unique to Jamie Otis. Take Julie Jeppson who runs the channel TheBigFamilyJewels on YouTube, where she shares videos showing what it's like to raise eight children. She started the channel because she was a stay-at-home Mormon mom with a husband who traveled often for his work as a musician (as of this writing, the couple is now separated). She was lonely and overwhelmed and she wanted to talk to other mothers. "It gave me this virtual community of friends, of other mothers who were dealing with the same thing I was, who were in the thick of motherhood and postpartum and loneliness and wondering if this was all we were meant for," Jeppson tells me, and my heart clenches because, same. She loved that vlogging was a way of creating memories that her kids would be able to look back on. "Nobody even has VCRs anymore. But being able to just get right online and just type in 'Ari's first grade performance' . . . and find it in minutes and then just laugh about it. So for me, it's the hustle and the side income that it brings in, but also being able to look back on past videos and just for memory's sake and the documentation of my own family history."

With 214,000 subscribers, Jeppson doesn't make enough money to entirely support her family of nine, but she's hoping that she'll be able to grow her channel to the point that she does. When she and her husband were still together, the YouTube gig was a fun side hustle, but now that she's the sole provider in her household, she is feeling the pressure of that role deeply. "I'm desperately trying to create more content and bring in more views and subscribers," she says, and there is part of me that respects her honesty.

She claims that because of the legal circumstances of Jeppson's separation from her husband, she's unable to share the details online, which is frustrating to her because she's sure that if she could tell the story of the dissolution of her marriage, "my views would be insane right now." But she says she can't, so she's left brainstorming—and facing backlash from detractors who say she's "disgusting" and a "walking advertisement" who will "do anything for views" (this kind of vitriol is commonly lobbed at family vloggers). The noise from the haters is fine by Jeppson. "I might as well vlog every now and then and get products and be able to get paid for promoting products while staying home with my kids who are heartbroken and depressed," she says. She even has these conversations with her own sister, whom Jeppson characterizes as super against social media. But it's Jeppson's livelihood. "I have to do things online right now to survive. I don't want to have to sit here and have to explain why I'm doing this. It's not just to put my kids in harm's way. I have to do this."

Jeppson says she spent sixteen years as a stay-at-home mom who took care of the kids and the house while her husband advanced his career. If she hadn't made those sacrifices, "I would have been a neurosurgeon by now." But the time passed, and she stayed home, and now she's not only a stay-at-home mom of eight but a single stay-at-home mom of eight. And what is a single stay-at-home mom of eight supposed to do for work?

I ask her what kind of content brings in the most views and garners the most interest. "The ones that you're not supposed to post," she answers simply. "What does that mean?" I ask.

This is why she gets nervous speaking to journalists, she says, before revealing that the content that performs the best is, basically, the videos where someone is hurt in some way. "The videos that got the most eyes on them are the ones that had the bloody noses, or the broken arms, or the emergency room visit, or whatever. Those ones would get a lot of eyeballs. It wasn't like we were trying to have those happen. It was just daily life. The amount of teeth that my kids were losing and the amount of broken arms and strep throat that we have had among eight kids and the barfs and the colds and all the things, that was my thumbnail every day. People would be like, 'Oh my gosh, your kids are always sick, and I swear that she's doing this on purpose and blah, blah blah.' They can think what they want, but when they're in a household with eight children under the age of ten, we're all sick all the time. Those would be getting a lot of views every time that we posted an injury or even going to the orthodontist and showing the retainer in someone's mouth. Because humans are curious, and they want to know what is happening and why is this kid in the hospital? So they click on it."

The click means money for Jeppson. But why does she think that people are so interested in content in which her kids are sick or hurt? What is the fascination? "Well," she says. "It's the same thing, I think, with movies. Lots of people are driven to violent movies. Human nature is curious. It's not something that you see every day. It's like, why else do people rubberneck? Because they're curious and they want to see what happened. I just think it's human nature to be curious and to want to know, well, why is her mouth bleeding? What happened? Why does she have to go to the ER? I broke my wrist, and my thumbnail was an emergency trip to the ER. Of course, a lot of people clicked on that because they wanted to know what happened."

Scrolling through Jeppson's most popular YouTube videos, seven of the top ten are related to sickness or injury like the second-most popular video, "Update on 4 Year Old Clubfoot Relapse—Clubfoot Journey," and the fourth, "Surgery for 4 Kids in One Day! Panic in the O R Live Look at Tonsils Up Close," and the eighth, "Big Brother Consoles Little Brother After Getting a Flu Shot." I appreciate Jeppson's honesty here, but I'm also disturbed by what it means for both her kids and the viewers tuning in to this content. When one of Jeppson's children is hurt or sick, is her first instinct to grab her phone and document it? And how does that change her children's experience of being taken care of by their mother? I'm also side-eyeing the viewers who crave this content; Jeppson wouldn't be making it if there weren't a demand. What does it say about viewers that this content is what performs best? Dr. Krysten Stein, a professor of communication at the University of Cincinnati Blue Ash College, says humans are just naturally drawn to depictions of suffering. "Voyeurism and spectacle are the two words that keep coming to mind. It catches people's attention." And, Dr. Stein says, when you spend more time on a video of a child getting hurt than on one of a healthy child, the algorithm learns what you're interested in and feeds you more of it.

On one Reddit thread, commenters whisper about the nerve of Jeppson to post a video showing her son groggy from anesthesia meds. I watch the video, which is called "12 Year Old Gets Emergency Surgery at Midnight Scary" and has 50,000 views, which does not seem like enough to possibly warrant filming all of this. About seven minutes into the video, Jeppson films her son Xander waking up from anesthesia after appendix surgery. The child looks loopy, confused, and bleary-eyed. At one point, he seems like he's trying to conduct an orchestra that isn't playing. "He says he feels drunk," Jeppson laughs into the camera and pans back to Xander. "Did I have some beer?" he asks with his eyes closed. "Is that why I feel drunk?" Xander explains that he was at the bar last night, which is why he feels like this. He speaks with his eyes closed and Jeppson asks if he feels okay. "What does 'okay' mean?" he asks.

Six months after this video was posted, Jeppson arranged for me to speak to a few of her older children. I ask Xander what he thinks about being featured in his family's social media accounts, and he says it's "a pretty good thing because, you know, people are looking up to me." With other family vloggers, you can tell their content is staged, Xander says, but not theirs. "Even though it's intense, it, like, teaches other people how a life of eight kids is." What does he think is the most intense thing he's been a part of on the channel? "Maybe one of the two times I've had surgery," he says. "That was pretty intense because I was like, you know, in intense pain and stuff like that. But at the same time, you know, it teaches kids what to do when they have surgery."

This is such a childish, rosy view of the content that it breaks my heart. I don't doubt that Xander thinks kids are watching videos of his surgery to prepare for their own, but I do doubt that's where most of the views are coming from. I think that, mostly, it's people who can't help but stare when a child is in pain.

Were there ever times when he didn't want to film? I ask.

"Maybe like, some parts of the surgery," he says, and I wonder which parts because it seems like viewers can watch it all. "Like I was just lying there in a lot of pain and I didn't want to be filmed. And I was like, 'Hey, Mom, you can't film. You can film all the other parts of this surgery, just not like, this part in specific,'" he explains. Right, I say. "It's obviously the way she gets money for eight kids and one mom," he says, unprompted. Right, I say again.

There's something else behind the interest too, Jeppson thinks, something softer. People care about her family and her kids. They're invested. They know what each of her kids' favorite color is and what food they like best and which sports they're participating in. They know the story of her family so when someone is hurt or sick, they want to know about that too. In a way, she thinks, it's only natural. And Jeppson knows that there are detractors who think she's a bad mother who's exploiting her kids but, she tells me, her kids love YouTube. They want to have their own YouTube channels and monetize them "so they don't have to work for an employer for five dollars an hour. But there's pros and cons to everything. Sometimes they'll get made fun of for something, but it's part of the beast in the industry, I guess."

Jeppson's framing makes me uneasy. She's not making enough money through her YouTube channel to completely support her family (though as a single mother of eight, I'm sure every dollar counts), and her kids are invested in the content creation, but they're also getting bullied for it. And she just considers that to be a consequence of the work? My mind ping-pongs even as I write this. What else is a single mother of eight supposed to do to make money? But then again, what do other single mothers do? And doesn't every kid get bullied for something at some point? But then again, why give the bullies ammo?

One of her kids did a performance during Halloween, Jeppson tells me, where he was dressed up as a zombie and singing and dancing, and Jeppson posted a video of it. "Some of the junior high kids were just making fun of him," she says. "Really, in my opinion, I think that they're just—I tell my kids anyway—that they're probably just jealous. Everybody wants a YouTube channel and you can just tell them, 'Thanks for the view,' because they got on there and watched it and their view made you money."

I can't help but wonder if her son really did say that to his bullies. It strikes me as a line that a kid would say in a movie. If he did say it, how did it go over? And how much money did their view actually net the family?

A video of a child doing a Halloween performance is relatively harmless in the grand scheme of things. It's silly, sure, and it can be used by bullies, but it's not achingly personal or wildly vulnerable. Not like the video Jeppson posted that featured her younger daughter crying over her father, from whom Jeppson had recently separated. The full video begins as one of her toddler sons puts on underwear over his shorts then it cuts to a comparison between her children's feet, one of whom has a club foot (which has also been made into lots of content.) The next clip shows the family arriving at church followed by one of her sons doing what she calls "an exotic dance." ("Hey Magic Mike, what you doin'?" she says in the video.) The subsequent clip is of a child's broken thumbnail being pushed away from the nail bed, and it's graphic enough that I cringe and look away until it changes. This disjointed format is par for the course for Jeppson's videos, which often hopscotch between topics and visuals, with graphics only appearing on screen for a few seconds at a time.

At two minutes and four seconds, the clips change again and Jeppson is shown in a drab beige room with her arm around a crying child. On the screen is the text: "Sorry it's so quiet. We were in church." "Poor thing didn't wanna go sing," Jeppson says into the camera while rubbing the shoulder of her crying daughter. "They're singing Father's Day songs in sacrament meeting." She pans the camera around to show the room they're in as her daughter India sobs. "We came in here so she could have a minute," Jeppson says, not looking away from the camera lens. She pulls the camera closer to India, framing her in its eye. "I haven't seen those sad eyes for awhile," Jeppson says. "Tell me why you didn't wanna sing. Just too hard?" India nods. "I wouldn't have anyone to sing to," she says and looks into the camera lens. Jeppson explains that India can think about the father figures in her life as opposed to her biological father and that there's always the Heavenly Father who will never leave her. India continues crying and mumbles something I can't quite make out about "before he left." Jeppson rubs her shoulder and looks into the camera the entire time. Through the emotional scene, Jeppson only talks to the camera's lens instead of directly to her heartbroken daughter. The entire sequence is uncomfortable to watch: Why do I have a front-row seat to this child's sadness? And why is her mother holding the camera instead of using both arms to hold her? When India is finished speaking, Jeppson addresses the viewers: "This is rare for her, to show emotions," she says. "She's usually pretty closed down about it."

After that pronouncement, there's an abrupt jump cut to another child coloring. A few seconds later, an errant thumbnail is shown being ripped away from its bed. "It won't hurt you," Jeppson promises. The last thirty seconds of the long video shows Jeppson wiping her own bloody nose with a diaper. "I might have to go to the ER again," she says. "This one won't stop." She pans the camera toward the bloodied diaper, rivulets dripping down her nose into her mouth. "Fun!" she says. "This one will get me a lot of views. Clickbait!"

This is how Jeppson describes the process of filming that video:

"India was crying before I pushed Record. I said, 'Do you want to record this?' She was like, 'Why do you think I should?' I said, 'Well, it's not up to me.' I said, 'If you want to film it, though, it's always good because number one'—and this is where people will disagree with me and that's fine, but number one, there's a lot of viewers that watch and will write in and say, 'Oh my gosh, thank you so much. You helped me to get through this day, or whatever.' But I was like, 'India, there are other girls that are dealing with what you are. Do you want to share what you're dealing with and see if it might help someone else?' Then number two, I said, 'Sometimes it's good to talk about your feelings and explain what's going on and then be able to look back on this a few months from now and see how far you've come.' But I always ask them first when it's a vulnerable moment like that."

I've read comments on Jeppson's videos that thank her for her unpolished content and describe how it's helped other people get through their own difficulties. I just don't understand how that is more important than allowing India to experience such an emotional moment privately. Sure, it might help other kids who are missing their dads, but in my mind, other kids aren't Jeppson's concern. India is. And I question whether the camera being brought out at that moment served her. How can a child understand the internet or grasp its permanence? "How does a child refuse consent to the very person they rely on for survival?" writes therapist Robyn Koslowitz in *Psychology Today*. "A child cannot give informed consent to being the subject of public scrutiny. They lack the executive function to understand the long-term implications and the emotional freedom to say, '*I don't want this.*'" This paradox was further explored in the paper "Parent Influencer Activities and the Involvement of Children," where researchers wrote, "Parents are considered the primary caregivers and the main [people] responsible to protect their children's rights, but at the same time they are the ones engaging in influencer sharenting potentially violating their children's rights."

India is one of the kids Jepsson arranges for me to speak with, and I ask her about the video of her crying in church. "I'm okay with it," she says. "Because all YouTubers have a good side and a reality side, like a happy side and a sad side. And like, everyone needs to know the truth." At eleven, India already has her own Instagram account (though the bio says it's managed by her mother) and she tells me she hopes she has her own YouTube channel with her own kids when she's grown up.

Jeppson tells me that she asks her kids' permission before posting videos like that one, and India said it was okay. She tells me that it seems like just because she doesn't film herself asking and receiving permission, no one believes she's actually done it. The tone of our conversation has shifted; she's prickly and defensive, and I get the feeling that I'm annoying her. "Is that frustrating to you?" I ask.

"Yeah, it is. It bugs me," she says. "I'm not telling them what to do, but they happen to be watching me. So if you're uncomfortable with what you're seeing, you don't have to watch this. I don't go watch horror movies because I'm uncomfortable [with them]."

After we hang up the phone, I sit there thinking: *Did she just compare family vlogging to horror movies?*

3

Privacy for Sale:

Monetizing Everyday Intimacies

The Family Fun Pack YouTube channel has more than 10 million subscribers. They've published nearly 3,000 videos that have been viewed over 15 billion times. There are eight kids in the family and Kristine, the mother, tells me she started vlogging in 2011 with a video of her twins. She didn't understand privacy settings at the time and meant to just upload the video to send to her mother-in-law, but before she knew it, the video had been viewed 8 million times. It's a charming video in which Kristine tells her twin toddler sons to go to bed and they run, pajama-clad and holding bottles of milk, into their room and catapult themselves into their cribs. "Night-night!" the boys say, giggling. The view count of the video is over 130 million. "Everything just spiraled from there," she says.

Around the time their channel reached one million subscribers, Kristine and her husband, Matt, considered YouTube as a serious career possibility, but they were worried. What about medical insurance? What about a 401(k)? They debated the pros and cons before Matt left his teaching job, Kristine says. But now, YouTube is the family business, and when they're out in public, they're often recognized by fans. "It's a blessing for sure," she says. Do they have a plan for how long the channel will last? Is there a point in the future where they can imagine stopping? "We don't have a stopping plan because I think that's unethical to just leave your fans in the dust when they've been following your life for . . . How long has it been? Thirteen to fourteen years?"

At the start of their YouTube journey, Kristine used pseudonyms for her kids. It was Kristine's way of protecting her children's privacy even as their collective star rose. She says to this day, the internet doesn't accurately know her children's birthdays, though eventually she phased out their pseudonyms and began using their real names. She makes efforts to conceal information like the children's schools or what soccer teams they play for. The backlash against family vloggers hasn't escaped her—she knows people worry about the privacy of kids like hers, but she's trying to mitigate the possible negative effects. "I wanted my kids to have the ability to be completely anonymous if they choose to when they're eighteen," she tells me.

How is she doing that when videos of them have been viewed over 15 billion times? Well, she says, no one knows their birthdays or middle names or even the family's last name. They're even registered at school under a fake last name, Kristine tells me, in an effort to shield them. What about all the photos and videos? Is there a way to be truly anonymous when your childhood and adolescence have been digitized publicly?

"Well, obviously, they're going to get recognized, and that's just part of life. Their videos and photos are definitely out there, but that is like, that's the life we lead. And I mean, obviously, if someone really wanted to find them, they'd probably be able to. But you can be pretty anonymous as well. You can make your profile picture of a flower, and no one's ever going to figure out who you are. I think the kids will be fine."

There is, of course, the possibility that Kristine's kids won't be interested in anonymity. To find out, I asked her if I could speak to one or two of them and she agreed to what turned into a round-robin call where the kids passed the phone amongst themselves and answered my questions before handing me off to the next sibling.

Alyssa, who is seventeen, goes first. When I ask her what it's like to be part of a YouTube family, she says she gets this question a lot. She's homeschooled, she says, "just because of our busy schedule . . . so we have a lot more flexibility, more time to film videos and all the stuff we have to do. Obviously, we film a lot. Not too much, though." Plus, she films and edits content for her personal YouTube channel, which has 1.3 million subscribers. She's often recognized in public with her siblings. "Usually, we'll notice [a kid] staring a lot. They'll probably tell their mom, 'I know them from YouTube.' They'll look, and then once they get their confidence, they'll come up to us and then they're like, 'Oh, my daughter watches your channel. Can she get a picture?' And then we're always happy to say yes, and we'll take a picture with them." By the way, she tells me abruptly, Zac and Owen are here too if I want to talk to them. I say I do.

Zac, fourteen, is handed the phone next. He likes being in a YouTube family, he tells me. It's work, but it's fun work. He likes doing challenges with his family (like one where the kids guess the flavor of Pringles they're eating in a video that has been viewed 51 million times) and he doesn't mind being recognized in public, as long as fans ask for photos instead of just taking them secretly. He has his own YouTube channel with his twin, Chris, that has over 731,000 subscribers. When he's older, he wants YouTube to be his full-time job just like his parents.

Zac hands the phone to his little brother Owen, seven, who also has his own YouTube channel with 61,000 subscribers. Owen likes knowing that so many people watch the family's videos, and his favorite video is the vlog of his birth. "Because it shows all the memories," he says. The vlog of Owen's birth has been viewed over 9.7 million times. It captures Kristine in labor editing a video, to which her husband says, "She never stops working." The video cuts from Kristine getting ready to push, to the jubilant scene of Owen being placed on her chest. Kristine grins and Owen cries and watching, I'm touched as the baby gets weighed and measured and swaddled, and then Kristine holds him close as a cinematic tear drips down her cheek. The video concludes with a still photo of Owen, over which the words "Be sure to watch the kids' reaction to meeting Owen!" are transposed. The prompt to click to the next video shakes me out of my reverie and reminds me what this is: a business.

David, who is seventeen and Alyssa's twin brother, speaks with me next. (His personal YouTube channel has over 270,000 subscribers.) When I ask him what it's like to be part of a YouTube family, he tells me it's "really cool. It's definitely unique. We get to travel a lot. We have a lot of cool experiences. We go to a lot of places." The experience of filming, he says, depends on the video. For his own smaller channel, the videos are more structured and scripted, but the main channel is vlogging, which are clips taken as the family goes about their day. Sometimes they film challenges, which takes planning. "We make a Google Sheet and plan out what we're going to do," he says. It can be frustrating to film with the entire family because of the difficulty of wrangling so many kids. "Sometimes you just are bored waiting while somebody else is trying to get the right clip, and it could take a little while. It's pretty boring and sometimes hard to film and get it right." David goes to online homeschool because "it's convenient and flexible," which is important due to the family's busy schedule. "We do a lot of traveling and events and stuff. And if I was in a public school, they don't like you missing. And if you miss too much, you won't graduate. You just can't miss too often. Whereas being online, I can do [whatever work] I miss when needed."

Chris, fourteen, takes the phone next and tells me he loves being on YouTube. One of his favorite videos was a challenge in which the kids spun a color wheel and had to eat only foods of the color they landed on for twenty-four hours. The video, which featured Chris landing on the color green, has been viewed 49 million times. In the future, Chris tells me, his goal is "probably to do YouTube. If I have a family, I'll do [family content]. And if I don't have a family yet, I might do stuff like pranks. I'm not sure."

One of the younger kids in the family, Chloe, is only two years old so obviously she can't speak to me, but I wonder what she would say if she could. Recently, Kristine posted a video called "Potty Train Your Child in 1 Week 10 Tips—Mom of 8!" featuring the details of Chloe's potty-training journey. The video opens with Chloe refusing to say hi to the camera—"she's sometimes a little shy," Kristine says before asking Chloe if she wants to go potty. Chloe takes off running and the camera follows her into the bathroom, zooming in on her tiny potty-training potty. There's a jump cut to the aftermath of Chloe going to the bathroom, where she's reaching for a Hershey's Kiss, which Kristine calls "peepee and poopoo treats." Chloe eats the Hershey's Kiss and jumps with excitement. Kristine and Chloe then go into Chloe's underwear drawer where Chloe exclaims over her undies. Chloe grabs a pair of underwear and runs over to her dad; Kristine turns to the camera to explain the potty-training journey of each of her children.

Honestly, the video is boring. I scroll through the timeline of the video, searching for the segment YouTube marks as "most replayed." "I'll tell you guys about one of Chloe's accidents that she did have," Kristine says, and I am not at all surprised that this is the most replayed part of this video. The story goes as follows: Kristine was out walking the dogs when her teenage daughter Alyssa texted her and said that Chloe had pooped, but they couldn't find it. Kristine asks all the kids to look around for the undies—they look under the piano, in the ball pit, in the trash. Suddenly, Kristine hears Zach say he found the undies—in his bed. "It was like artwork all over the wall and the bed and the stairs," Kristine's husband offers with a grimace. "But what you need to do when you have an accident is show your sheer and utter disappointment," Kristine says. "She feels horrible if I say the words *oh no* in a deep voice. I just kept saying 'I can't believe you pooped in Zach's bed, you ruined your undies, I can't believe you did that.' You just keep letting her know you're so sad. And they feel guilty, they feel horrible, they realize they've done something wrong and that just reinforces, you know, where does the poopoo go? It goes in the toilet." The top-rated comment on the video reads "Chloe has grown so fast love you guys."

So I couldn't talk to Chloe. But I could watch her get potty trained and hear about the missteps she had along the way. (Sam [a pseudonym], the parent of a formerly wildly popular vlogging family, whom we'll hear more from in a later chapter, tells me the potty-training video doesn't surprise them. "Few people stop to think about how having themselves as a baby advertising their potty training all over the internet will affect those kids when they get older. We certainly didn't think about it at the time," Sam texted me. "All we thought about was $$$. It wasn't uncommon for us to get between $10k to $25k for one sponsored 1.5- or 3-minute segment in our vlog. Creators bigger than us were bringing in $50k to $100k for a similar brand deal.")

The responses I've gathered from the older kids of the Family Fun Pack make me wonder: *Is all this hand-wringing about the privacy and rights of YouTube children useless? Especially if the children enjoy it?*

It's important to examine the privacy and monetary implications of a life lived online, but at the same time, we cannot ignore that some of these kids like being YouTube stars. Of course they do! They have fans, they make money, and they're living the dreams of the 86 percent of young Americans who want to be influencers. Amy Golden, a high school career counselor in Los Angeles, told me, "I ask the kids what they want to be in the future. And every single one of them wants to be an influencer. That's all they want to do." If that comes with the reality of knowing (and accepting) there are videos of them crying as toddlers on the internet forever, does it really matter? It doesn't seem like it does to the Family Fun Pack kids at least. It's not useful to categorize every instance of a kid growing up on YouTube as dangerous exploitation that will incite resentment and feuds in the future. That kind of binary thinking fails to explore the intricacies of this world. Kristine, for her part, is insistent that her children are living the dream. "Every kid would love to be on YouTube. You know what I mean?" Kristine says. "Most kids are like, 'Oh man, I'm jealous,' or 'My mom won't let me start a channel,' stuff like that. So the kids know that they're kinda lucky that they get to do it."

While studying for a master's degree in bioethics, technology, ethics, and policy from Duke University, Bridie Hamilton wrote her thesis "Anything for Views Parenting: Framing Privacy, Ethics, and Norms for Children of Influencers on YouTube." After studying videos of prominent YouTube family vloggers, Hamilton was able to outline what she calls the "salient characteristics of privacy violations."

The first harm Hamilton notes is "despite an account being registered in a parent or family name, a child or children is disproportionately the focus of the content." Most family accounts, even when they're registered under the parent's or family's name, focus on the children. When people talk about the mom influencers and family vloggers they're fans of, they talk about the kids, not the parents—because the kids are the stars of the show.

Second, Hamilton writes, "Filming occurs regularly for prolonged periods, often on a predictable schedule." Though we can't know how long filming takes for these accounts, the amount of content they're creating suggests filming lasts for prolonged periods.

The third item on Hamilton's list of what she perceives as privacy violations is "work [being] indistinguishable from play." When I read this, I think of Iris, the nanny of the child of an influencer, who said the young boy she nannied for could feel but not understand the difference between playing in front of the camera for content purposes and playing just for play.

The fourth item—"Emotionally volatile or misleading content titles (clickbait)"—are the bread and butter of family vlogger parents. Consider "Our First Pet Died": a video posted by the LaBrant Family (who have 12.8 million YouTube subscribers) that has racked up 34 million views. The thumbnail is of their daughter Everleigh, then around five years old, crying and hugging her mother, Savannah. At the time of the video, the LaBrants had a tiny dog named Carl (the same dog they once told Everleigh had run away as a prank), and I'm sure that viewers, seeing the thumbnail of Everleigh seemingly overcome with grief, must have thought Carl was the one that died. The video reveals that the deceased animal was actually Clever, a goldfish. When Everleigh's stepdad Cole tells Everleigh and Savannah the news, they take it in stride and head to the backyard to bury the fish. After the burial, the trio head to the pet store to buy a new pet. The thumbnail image of Everleigh crying into her mother's hair never appears in the video.

The next listed characteristic of possible harm—"homeschooling, frequent relocating"—seems ubiquitous among vlogging families. Although there aren't currently statistics on how many family vloggers homeschool their kids, it seems like many, if not most do (with critics accusing the vloggers of homeschooling in order to have more time to create content).

Next is "revealing personally identifiable information." I think of how kids born into influencer families immediately have their birth date, first, middle, and last names, and measurements shared with the world, violating Hamilton's perceived harm of "revealing personally identifiable information." Take, for example, Christian Anthony Burrello, 9 pounds and 6 ounces, born on Friday, August 23, 2024, at 7:08 p.m. to mother Caila, who has 488,000 followers on Instagram.

Next is "frequent disclosure of the child's physical location," which brings to mind mom influencer Jamie Otis, who has one million followers on Instagram and tagged the hospital where her newborn son was receiving medical care.

The next item on Hamilton's list of characteristics of possible harm—"revealing intimate details about the child's identity, developmental, or intellectual capacity"—brings to mind the plethora of mom influencers who share details of their children's medical diagnoses. Caitlin Nichols, an influencer with 795,100 followers on TikTok, posts videos of her triplet daughters who were born prematurely. The words "I am so tired" captioned a video she once posted featuring one daughter asleep on a hospital bed wearing a breathing mask.

"Filming embarrassing, personal, confidential, or intimate conversations" is something vloggers do in video after video. I think back to the Family Fun Pack filming the children's grandmother's funeral and their reactions to it in a video viewed 1.9 million times and hashtagged #familyfunpack and #grief.

Hamilton lists eight possible harms to children of influencers: (1) unwanted attention from invisible audiences, (2) risks to personal safety, (3) loss of control over personal information, (4) stigma, (5) psychological consequences, (6) reputational damage, (7) loss of discretion or seclusion, and (8) unknowable harms. It's the unknowable harms that really worry me.

Emily Kline, a clinical psychologist and writer, worries about the dynamic between influencer kids and their parents. "There's that concern of being able to trust your parents, to keep their confidence in them, which I think is really important," she said. I think about a former influencer kid who I interviewed for a story who told me they stopped telling their parents things because they were worried the disclosures would be turned into content. "Then their parents are like, 'Why doesn't my kid tell me anything?'" Dr. Kline says. "And it's like—because you're turning it into content."

Consider Rachel, the teenage child of an influencer who stopped inviting her friends over to her house because she worried they would be captured in her mother's uploads. Sometimes Rachel's mother asks her to make videos and assures Rachel that it's not going to be posted online, that she's only going to send it to Rachel's cousins or aunts—but then when Rachel checks her mother's Facebook, she spots the video. "There were things she told me she wasn't going to post. I'm like, 'Mom, please don't post this,' and she'll still post it. And I started crying because I just really didn't want that. That was private to me . . . and now it's for her [followers]." I ask Rachel what her mother did when she started crying after finding the dancing video posted on Facebook. "She didn't see [me crying] because she was in the car editing videos," she said.

In contrast, there are the child influencers who have now grown up and grown into their own careers as influencers. Take Brooklyn and Bailey McKnight (@brooklynandbailey on Instagram, where they have 9.6 million followers; TikTok where they have 7.3 million followers; and YouTube, where they have 7.3 million subscribers), who are, of course, Mormon. (More on why so many influencers are Mormon in a later chapter.) The McKnight twins are now twenty-five years old, but they first rose to fame when they were featured in their mom's YouTube videos at around nine years old. Mindy McKnight, who has 5.7 million YouTube subscribers, created videos featuring herself doing her twin daughters' hair in enviable styles. (Though Mindy's most popular video by over 110 million views shows her daughters swimming in mermaid tails and being sprayed with water.) By the time Brooklyn and Bailey were thirteen, they had their own channel, and by fifteen, they were recognized by *Business Insider* as one of "13 up-and-coming

YouTube stars you should be following." In 2026, they have their own skin care line (@stayitk, which has over 124,000 followers on Instagram) and a clothing line (@lashnextdoor—256,000 Instagram followers).

Brooklyn and Bailey were raised on YouTube, and they've turned their childhood influencerdom into sprawling careers. I follow the twins, and I'm endlessly impressed by the way they've transitioned their audiences through their life phases. The description of their YouTube channel reads: "We are two twin sisters who just vlog their life and share it for y'all to enjoy! From wedding content, building houses, college, friendships, periods, cooking/baking, challenges, vlogs, travel, dance, and SO MUCH MORE, we film it all. Stay, subscribe, and enjoy." Their first video was posted eleven years ago and has 3.1 million views. In the video, thirteen-year-old Brooklyn and Bailey have braces crisscrossing their teeth as they participate in a Q&A. They explain how to tell the difference between them. Bailey, they say, has a more squished face because of how they fit in their mom's womb, while Brooklyn has a rounder face. Bailey wears long earrings, but those are "not part of Brooklyn's personality." Bailey has bangs. The list goes on. Their favorite food is potatoes ("all kinds of potatoes"), and Brooklyn loves sci-fi and fantasy books while Bailey's favorite is the *Goose Girl* in the Books of Bayern series. Two years later, Brooklyn and Bailey post a "Get Ready with Us: HOMECOMING 2015" video. The girls have their braces off and they both have dates ("They're just friends," Bailey explains in the intro). The entire week of homecoming festivities is filmed, and the video culminates with details of the girls' homecoming outfits and a clip of the dance itself.

Brooklyn and Bailey's YouTube channel is a veritable scrapbook of their lives from the age of thirteen to the present. They debate where to go to college (they eventually decided on Baylor University, which they announced in a paid partnership Instagram post, where they said

Baylor was "welcoming, Christian, had challenging academics, and was close to home"), reveal boyfriends and go on dates, and eventually get engaged and married (Bailey first, to her high school sweetheart Asa, followed a year later by Brooklyn, to Dakota, who was number 7 in "Brooklyn's 10 Dates in 10 Days" YouTube challenge). Their content has changed with their age, like when twenty-two-year-old Bailey got married and started selling vibrators on Instagram, presented with the messaging that female pleasure is important and should be prioritized.

These girls—now women—have grown up on YouTube, and it doesn't seem to have scarred them. Their public, online childhood helped them build lives that appear to include financial stability and freedom. Brooklyn and Bailey exemplify something that we can't ignore: there are influencer children who, far from hating their parents for putting them online, used the social media fame they were raised with to shape their lives. And they seem like nice lives—the twins own businesses and houses, and I can't even imagine how much money they have in their savings accounts. Brooklyn recently gave birth to her first child, a son named Archer. Although she makes content around motherhood—like a baby shower haul or Archer's birth vlog (viewed 3 million times)—Brooklyn has never shown her son's face online. When she posts him, she covers his face with an emoji (most often, a brown heart) or angles the camera so his face is hidden. As a child influencer who has grown up online, Brooklyn is making a starkly different choice from her parents with her own son and his online footprint. In the comments of her posts, fans bet when there will be an "Archer face reveal" and hypothesize about whether he looks more like Brooklyn or her husband, Dakota.

Avia Colette Butler is another child of influencers turned influencer. She is the daughter of Shay Butler, the creator of the @Shaytards YouTube channel, one of the first family vlogging channels. (The most popular video on the channel, with 30 million views, features Avia and is titled "Babytard Is Sick!" It details a hospital visit for a stomachache and fever that results in no diagnosis.) Avia, now nineteen, was only three years old when her dad uploaded his first YouTube video, which garnered an impressive 2 million views. Now, Avia has 519,000 followers on TikTok (where her bio reads "luckiest girl in the world" followed by an email address for her representation and a linktree that connects users to her other accounts), 280,000 subscribers on YouTube (bio: "Follow your dreams. I dare you"), and 242,000 followers on Instagram (bio: "Just me slaying that's all"). In a 2023 TikTok video, Avia does her makeup while responding to a comment that says, "Can you talk about your experience as a child of a family vlogger?"

"It's sad to say that there are a lot of sad stories that pertain to family vlogging of kids getting screwed over by their parents and just all of those sad things," Avia says in the video. "And it makes me so sad that some people had to go through that, that some kids had to go through that. But me, on the other hand, fortunately? That was the best time of my life." Avia delicately dots contour on her cheekbones as she says she was "living in the good old days without even knowing I was living in the good old days." She loved when her dad pulled out the camera and being a "little influencer," even though she didn't even know what an influencer was and that word wasn't yet used to describe their job. "I also liked it because I will go back and watch old YouTube videos of things that happened to me that I don't remember, because you're not gonna remember every single day of your childhood," she says. "I'm just so grateful [my parents] did it because now I have my job today because of it and I'm forever grateful for that."

Avia, who wants to be an actress, has moved to Los Angeles to pursue her dream. Her social media fame seemingly finances both the move and her daily life so that Avia gets to focus on trying to be an actress (while living in a large, airy apartment and wearing beautiful, expensive clothes). I think again of her TikTok bio: *luckiest girl in the world.*

Everleigh LaBrant is eleven years old, and she has 5 million followers on Instagram. Most intimate details of her life are a Google search away: She was born when her mother, Savannah, was nineteen years old; her biological father died in 2022; she calls her maternal grandmother Gigi. You can watch her grow from a baby to a gap-toothed toddler to a long-legged kid. There are videos of her dancing at dance competitions and playing with her (many) younger siblings and crying after her parents pranked her that she'd have to give away her puppy. There is an entire snark community on Reddit (r/LaBrantFamSnark) focused on her family with 45,000 subscribers. There are fan edits of her on TikTok that imagine her inner dialogue and slideshows on Instagram that have watermarks from the fans who created them.

Everleigh, in short, is a character—and a popular one. In the comments of her photos, fans bemoan how grown up she has become, how they've been watching her forever. "She has been gorgeous since she was a baby," reads one. "So beautiful, Ev! I love u always," reads one by a user with the account name @everleighlabrant.forever. "[T]he eye color is to die for," reads another.

In the world of child influencers, she's in a class unto herself. Her career started when Everleigh's mom Savannah, a thin blonde from Orange County, started posting mommy-and-me photos of the two of them. "There's just something about our brains. We love seeing moms look like their daughters," says Kelsey Weekman, a culture reporter at Yahoo News. "It's charming. It gets your attention. It's cute." Savannah was almost immediately popular online, and it's not hard to see why. "Some people are just born to be influencers," Weekman says. "Savannah had that thing." She was a pretty, young mom, only nineteen when she had Everleigh and when she posted photos dressing her toddler daughter in the latest fashions, people loved it. It didn't hurt that Savannah herself was the exact kind of woman who finds success on social media: white, thin, blond, and conventionally attractive. It was kind of a fairy tale too—the young single mom with her sweet daughter who looked just like her.

When Savannah was a twenty-three-year-old YouTube and Instagram star, she met nineteen-year-old Cole LaBrant, then a Vine star, part of the comedy trio Dem White Boyz, and the pair ascended to true social media royalty status. In a YouTube video titled "How We Met | Cole & Savannah," which has over 3.3 million views, Savannah and Cole describe, well, how they met. On the now-defunct app Musical.ly, Cole sent Savannah a DM asking her to share his content. "She's like one of the biggest people on Musical.ly and I just got on Musical.ly so I thought, *Hey, maybe she'd want to do shoutout for shoutout with me*," Cole explained in the video as he drove down a nondescript highway. "I heard her and her daughter were cute and funny." The story, as they explained it, was pretty boring, an explanation of their missed direct messages and viewing each other's pages. Things heated up when Cole, on a content trip to California, went to the famed shopping center The Grove and ran into Savannah and Everleigh. They arranged to meet at the beach and have dinner, spending hours sharing their life stories with each other. Before Cole flew back to Alabama, he kissed Savannah, "sealing the deal" with "my first kiss," he said.

Within a year, Cole and Savannah were married. Their July 2017 wedding was memorialized in a YouTube video that has been viewed more than 50 million times. In the video, Cole and Savannah both teared up as they delivered heartfelt vows to each other, and four-year-old Everleigh, in a white dress that matched Savannah's, grinned up at her mother and new stepfather. In 2018, Savannah became pregnant with the couple's first daughter, Posie Rayne LaBrant. They documented the pregnancy on an Instagram account with hundreds of thousands of followers that inexplicably has since been deleted, though the couple's other children still have Instagram accounts. On YouTube, Posie's birth vlog has over 24 million views and features Savannah pushing as Cole holds the camera. We watch them meet their daughter, covered in vernix, for the first time. Savannah and Cole immediately started posting Posie on their Instagram accounts as well as her own. "My family says I'm sent from heaven," reads one caption from the now-defunct account. "Because of you, I'm convinced every angel in Heaven is actually a bunch of babies singing around the throne of God (praise hand emoji)," Cole commented on the photo. Savannah, for her part, commented "We love you little angel baby" even though one of them was presumably responsible for the post in the first place. But this is the world of family influencers—you post photos as your infant and then comment back to them from your own account.

In 2020, Cole and Savannah welcomed their son Zealand Cole LaBrant, who has nearly 600,000 followers on Instagram as of this writing. Zealand's first post is a photo of him as a newborn, perfectly swaddled in a beige blanket. The caption reads "Started from the womb, now I'm here [sunglasses emoji]." In 2022, Zealand got a little sister—Sunday Savannah LaBrant, who was relegated to sharing an Instagram account with big sister Posie (the @ went from posie.labrant to posie.sunday before being taken offline). And in 2024, the youngest LaBrant was born. Beckham Blue doesn't seem to have his own Instagram account. Birth vlogs were posted for Posie, Sunday, and Zealand but not Beckham. After Beckham's birth, the LaBrants announced that Cole was having a vasectomy (and then they filmed a YouTube video about it that netted a somewhat paltry 730,000 views). "Everleigh's prayers have been answered," reads one of the top comments on the vasectomy video. There is a popular theory online that Everleigh has to act as more of a second mom to her younger siblings than a big sister, aided partly by comments Savannah has made about homeschooling Everleigh so she can help with the younger kids. "FREE EVERLEIGH" is a common comment under posts of the preteen. "Can't wait to hear Everleigh's story" is another.

In 2022, Everleigh's biological father, Tommy Smith, died. Smith's passing was a tragedy, and Savannah tactfully addressed it on Instagram, posting a photo of Everleigh and her dad with the caption "Our hearts are incredibly heavy as we process the loss of Everleigh's dad, Tommy. He loved Everleigh immensely. As we navigate through this difficult time we kindly ask for privacy so that our family may continue to love on Ev, pray and grieve with her. Your prayers for Everleigh are so very appreciated. [whiteheartemoji]." The cause of his death remains private, but internet speculation is rampant. The subreddit r/LaBrantFamSnark lit up with theories on Smith's death and just how Savannah would monetize it (she didn't). People combed through legal records trying to find autopsy results and posted old photos of Smith and Everleigh like they were mourning him themselves. They wondered how Cole was reacting, if he was glad to have Smith out of the way so he could have his perfect family. On TikTok and Instagram, fans posted collages and reels of photos of Smith and Everleigh set to heartbreaking music. For Everleigh, a ten-year-old who has been an influencer since she was two, even her father's death was transformed into content for public consumption. Not even this most intimate moment was shielded from the prying eyes of millions of fans.

What is it about Everleigh and her parents that is so fascinating? For one thing, they're just *so blond.* Like almost impossibly blond. They all have bright blue eyes and in photos and videos, they are nearly always color-coordinated and matching. They love Jesus, which they take pains to include in most of their posts. They recently moved from California to Tennessee because, they say, that's where God was calling their family. And their obsession with babies and pregnancy ("Savannah's either been pregnant or breastfeeding a newborn for 95 percent of our marriage," Cole once posted on Instagram, "I hope I still like her when shes not pregnant/breastfeeding") coupled with the way they talk about their sex life ("no longer a virgin," Cole tweeted on their wedding night, "being married rocks!") is something that's hard to look away from. But I think one of the most important reasons people are so fascinated by this family—and one of the reasons I am—is because Everleigh is the prototypical influencer child. She has been online since she was a toddler, and hordes of fans have watched her grow up. When people chomp at the bit for her tell-all or to hear what it's like behind the scenes, I think what they're really saying is this: Tell us if it was really as shiny as it looked. But I wonder if there's a part of us that hopes it wasn't. Do we want Everleigh to have a terrible story to tell? And is our anticipated consumption of it part of the problem? Dr. Carolina Are, an innovation fellow at Northumbria University, says one of the reasons that we anticipate the fall of influencers is because their lives are *just* out of reach. "The downfall is more enjoyable because it feels like, 'Oh they are just like me, but they made it,'" Dr. Are says. "A lot of people resent the narrative of making it in the influencing world. People resent that it's become a real job for others and they enjoy the downfall. It's almost like a socially mediated form of reality TV. People watch it waiting for the mess."

It makes sense that people are invested in Everleigh's story. We've watched her grow up, from those first YouTube videos and Instagram posts of her as a two-year-old. And some influencer kids, including her siblings, are making their debuts even earlier—like before they're even born. The Instagram account of Alessi Luyendyk, the six-year-old daughter of former *Bachelor* star Arie Luyendyk and his wife Lauren Luyendyk, was made when Alessi was in utero. (Which isn't that rare: A 2017 study found that 25 percent of fetuses have a social media presence before they're born, as a result of parents sharing sonogram pictures online.) Lauren and Arie posted photos of Lauren's pregnancy with captions written from their unborn daughter's point of view. "I'm 2/3rds of the way to meeting all of you!" reads one. "I just grew a bit and practiced my breathing movements to get my little lungs ready for the big day." Alessi's younger siblings, twins Lux and Senna, share an Instagram account that has 247,000 followers, though the captions for their in-utero posts were formatted as letters from their parents to the twins. (As of this writing, Lauren recently gave birth to the couple's fourth child, Livvy Rowe Luyendyk, after Arie's vasectomy was reversed, a process that was captured on YouTube.) In a podcast appearance on *Off the Vine*, Arie and Lauren talked about Alessi, then a preschooler, being recognized in public. "Even at her school . . . it was super weird because I would drop her off for school, and there was this one dad in particular . . . he would be like, 'Oh my God, is that Alessi? Is that Alessi?' All the moms were talking about her," Lauren said. "And she's like, visibly confused," Arie says. "That feels a little bit like crossing a boundary to me," says host Kaitlyn Bristowe.

I get it—it is weird for that dad to be talking about Alessi and it must be strange for her, as a child, to see other parents talking about her and staring at her. But listening to the podcast where Lauren and Arie talk about this happening, I also can't help but wonder what they thought was going to happen. Alessi has been made into social media content since before she was born, and now she's a tiny person living in the world—a world where people have been following her journey since utero. I don't think you can have both things—a platform built on the exposure of your child and privacy for them in the real world. Those two paths seem mutually exclusive. And as she gets older, I wonder what it will be like with her peers who may hear from their parents about Alessi or who can use their own internet access to see the details of her life.

Not every child influencer's content is centered around their personal lives. Take Ryan Kaji, the star of the Ryan's World YouTube channel, which has an astounding 40 million subscribers. Ryan became the most popular kid on YouTube by posting toy reviews and unboxing videos. In an interview with *Rolling Stone*, when asked about his earliest memory, Ryan says, "I remember I had a tablet. I was just clicking, scrolling, watching videos. I was a little kid, maybe three years old." In his first YouTube video, which has now been viewed 54 million times, three-year-old Ryan chooses a Lego "choochoo train" from Target. Now twelve years old, Ryan is among the most famous kids on the internet, with a television show, a feature-length movie, and merchandise valued at more than $250 million in 2021.

Though Ryan is arguably the most popular kid influencer in the world, his content is stunningly impersonal. In each video, he simply unboxes or plays with a toy. There doesn't seem to be the intimate aspect to his content that there is for other influencers. Scrolling through his videos, you don't see Ryan throwing tantrums or being disciplined or crying. His content is shiny and geared toward the ever-revered capitalist pursuit of selling more stuff. Even though Ryan is the most popular kid on YouTube, he seems to enjoy a level of personal privacy that other child influencers don't. You may have watched him for years, but there's only so much you can know about him.

This isn't the case with most kid influencers. We watch their breakdowns, triumphs, tears, potty-training journeys, and their efforts to learn how to tie their shoes and write their names. We watch them grow up. But on the internet, with a quick Google search or scroll through their parents' YouTube accounts, they are children forever.

I wonder about the stories that are going to follow them as they grow. That's something Bekah Martinez, a mom influencer with approximately 860,000 followers on Instagram with the handle @bekah, thinks about too. Bekah was a *Bachelor* contestant (actually on dad influencer and former Bachelor Arie Luyendyk's season) who gained popularity on Instagram after appearing on the show, which, she tells me, was the whole point of appearing on it. "When I went on *The Bachelor*, I told my dad, 'Hey, I'm going to be able to quit my nanny job after I go on this show if I go far enough on the show. I'm going to get enough followers, and [then] I'm going to get to do ads on Instagram,'" Bekah says. "My dad was like, 'What?'" If you're surprised by Bekah's candor, it's kind of her thing. Her Instagram posts and podcast appearances routinely make waves for her lack of filter, whether she's talking about having sex while breastfeeding her child (something she's routinely panned for, even years later when anonymous Redditors stumble upon the story and resurface it) or refusing to shave her legs and armpits.

After she was eliminated from *The Bachelor*, Bekah unexpectedly got pregnant only three months into a relationship with boyfriend Grayston Leonard. "You're like, *Who am I?* Because things have definitely changed. But how much of me has changed? What has stayed the same?" As Bekah found her footing within motherhood, she settled into the art of sharing it online—and what she noticed was that the more authentic her content was, the more people responded to it.

But sometimes, Bekah tells me, she stepped over from sharing into oversharing, like discussing the details of her sex life (which she's really trying not to do anymore). There are things that she regrets sharing. Once, during the recording of a podcast episode, Bekah talked to her cohost about how, when her daughter Ruth was teething, she opened one of Bekah's drawers and found a silicone vibrator and started chewing on it. "And I was like, 'Ha ha, whatever, that was funny.' And people definitely took that and were like, 'That's really gross and you're being perverted,' as if I'm intentionally handing my child a sex toy."

The story follows her; when Bekah is a guest on other podcasts, sometimes the vibrator story comes up and the backlash ignites once again. "It's like, 'Don't give this girl a platform. She's weird. She gets her kids sex toys to chew on while they're teething.'" She's tried to change how she shares about her kids. "For instance, if I'm asking for advice, I wouldn't say now, '[my son] Franklin keeps wetting the bed. Does anybody have something to help with that?' I would probably just say, 'One of my kids is struggling with wetting the bed.' So there's these frames that I started doing just to be mindful."

Bekah knows about the backlash against family vloggers and mom influencers—the idea that they're moneygrubbing child exploiters who don't want to get real jobs—but she finds that analysis too simplistic. "People come at me and they're like, 'How do you feel about posting your kids every second of the day or sharing so much of your children?'" Bekah says. "I guess, from my perspective, I share very little. I usually share one- to two-minute snippets from our days. I do not share ninety-five percent of our life." (Based on my following of Bekah, this seems true.) And she finds true purpose in her social media career. "I feel a strong conviction in the sense that I am supposed to be online," she says. "The way that I also see it is there's a lot of things as a child that you're not able to consent to that you might have issues with parents about later. There are all these different things, these choices that we make for our kids all the time that they might look back on and be pissed with us about. And the way that I share my children is, I think—if they grow up and they're really pissed that I had their face out there and was sharing them . . . that's something I'm going to have to take responsibility for."

But for now, being a full-time mom influencer has changed her life and the lives of her children. Bekah bought a house for her family, and she's able to stay at home with her kids. Does it feel like a fair trade, sharing some intimacies of their everyday lives in exchange for giving her kids a financially stable life and a stay-at-home mom? "Yeah," she says. "The way that I personally view my relationship with sharing my kids online, I do think it feels fair. And if they disagree with that in the future, I am more than willing to reckon with that."

On one hand, I think this is all a mother can do—make choices for her children and hope she won't be hated for them later. I truly believe in the value of women sharing their stories of motherhood and how overwhelming, difficult, and all-encompassing it is. It's saved me more than a few times in my own postpartum period to be able to read the experiences of other moms. I read through blogs that detail what that first month at home is like and sigh in relief that I'm not crazy for crying for an hour each day. I hungrily listen to audiobooks about motherhood while rocking my baby to sleep and feel myself tearing up when I hear a sentence that perfectly mirrors my own anxieties back to me. I scroll through TikTok and watch videos where moms put their babies to sleep, hoping to pick up tips to help my daughter sleep more than 3.5 hours at a time. It can't be overstated how much other mothers sharing their experiences has helped me through my own first foggy days of motherhood. And it does sometimes feel hypocritical for me to be writing a book about what it's like to be the child of a mom influencer or family vlogger when some of them have acted as guiding lights for me along my own path. I do think there is a way to both allow women to share their experiences of motherhood and parenting while simultaneously protecting the privacy of the children involved. And there's a difference between being the child of a mom influencer who sometimes shares stories of your bedtime routine and being the child of a family vlogger who is used to living with a camera in their face. (Isn't there?)

4

When the Cameras Don't Stop:

Meltdowns, Mistakes, and Crossing Moral Lines

"**Come closer for the video,"** Jordan Cheyenne coached her crying son, Christian. "Come closer. Put your head right here. Act like you're crying really quick."

"I am crying!" her son said between heaving breaths.

"Go like this," Cheyenne said, screwing up her face in a facsimile of a cry.

"No, Mom, I'm actually seriously crying," Christian insisted.

In the video, titled "We Are Heartbroken," Cheyenne announced that the family's puppy had a potentially deadly illness. As eight-year-old Christian cried, Cheyenne could be seen telling him how to place his hand and where to turn his face. "Let them see your mouth," she tells her son. "Look at the camera."

Cheyenne didn't mean to upload the video of her coaching her crying son to her 500,000 YouTube subscribers, but when she accidentally did, the uproar was immediate. The video validated the darkest concerns people had about family vloggers: They were exploiting their children's most vulnerable moments in front of cameras for views and followers. Watching the video of Cheyenne coaching her son to cry while he's already crying is so painful I want to close my eyes. It—like so much of the content family vloggers post—feels like a total violation for me to be watching a moment this intimate. A child is crying over his sick puppy—why is there a camera in his face?

In an interview with *Today Parents* after the outcry, Cheyenne said she regretted her actions but that they were par for the course for creator parents. "People will have their kids ham it up. Behind the scenes they're like, 'Do this, and I'll give you a treat.' It's so wrong and I can't even say how disappointed in myself I am," she said, adding that she had deleted her YouTube channel and gotten both herself and her son into counseling; but when I check, three years after the scandal, her YouTube channel is still active and boasts 504,000 subscribers though she rarely posts and gets low views. Her latest video is titled "How I manifested my dream life being DELUSIONAL (Here's what I did so you can copy me)." The video has only 3,800 views. Cheyenne doesn't seem to have been able to bounce back from her scandal—which internet culture journalist Kat Tenbarge explains to me, may be for a few reasons. First, though Cheyenne had half a million YouTube subscribers, she wasn't a *huge* creator with millions backing her—and the bigger the creator, the more they're able to withstand controversy. "If she had been a much bigger influencer, she probably would have kept going," Tenbarge says.

Take, for example, Colleen Ballinger (who sometimes goes by her alter ego Miranda Sings), who amassed tens of millions of YouTube subscribers through satirical videos and family vlogging. *Rolling Stone* published an investigation that alleged that Ballinger asked a minor fan about their virginity status and favorite sexual position; Ballinger did not respond to *Rolling Stone*'s request for comment. Though she has stopped posting on YouTube, she still posts regularly on Instagram to her 7.4 million followers.

There's also the Shaytards YouTube channel, which is headed by father and creator Shay Butler and often pointed to as one of the first family vlogging channels, which at its peak had over 5 million subscribers. The Butlers posted videos of the daily lives of their large, devoutly Mormon family—until February of 2017, when Shay was accused of cheating on his wife with an adult webcam star. Shay confirmed the allegations were true, telling fans on Twitter that he was suffering from alcohol abuse and planned to leave YouTube to seek treatment. Just nine months after the scandal, Shay returned to YouTube with a video in which he vowed to "take full responsibility. I don't expect any of you to forgive me. I just want to start fresh and not feel like a scum." Shay's wife, Colette, posted her own vlog in which she said, "I love my kids and I love Shay and we're working through it. We're doing the best we can, taking it one day at a time, and I can't wait to share it with you." Despite the fact that Shay Butler's entire YouTube kingdom was predicated on his reputation as a family man, the family's considerable online presence was able to withstand the cheating scandal. When an influencer who has millions of followers or subscribers—like Colleen Ballinger or Shay Butler—faces a scandal, there's just more padding for them to withstand the blow. Even if scandals or missteps may hurt them in the short term—in the form of losing some followers or sponsorships—they're still able to make money from YouTube AdSense or TikTok views.

Tenbarge also points to a personality trait that more successful influencers tend to have—the ability to persevere through controversies and negative attention. "It depends on your willingness to ride out that controversy," she says. "If you are super sensitive to online backlash, it will be harder for you to have the persistence to keep going as a creator." There's a world in which, if Cheyenne had just ignored the controversy or posted through it, she could still be a successful creator today—but she caved, and her YouTube house of cards fell around her.

Scandals within the family vlogging world happen frequently, shining a light on the ethical issues of the industry. What will viewers forgive? And what is a bridge too far?

Myka and James Stauffer, who documented their lives on YouTube channels to a combined 1 million subscribers, posted videos chronicling the steps of their journey to parenthood, including "Live Pregnancy Test! Am I Pregnant?!!!" and "My Miscarriage Story at 6 Weeks Pregnant!!!" In a viral story in *The Cut*, reporter Caitlin Moscatello detailed the Stauffers' road to adoption; in 2016, the couple posted a video promising to take viewers along on their adoption "journey." "I am so excited to open our hearts and see what the Good Lord has in store for us," Myka said. They narrowed their adoption search to China. As Moscatello wrote in *The Cut*, "China still accounts for more adoptions to the US than any other country, but now almost all adoptees from China to the US are toddler age or older, and many have existing health conditions. If the Stauffers adopted from China, they would almost certainly be choosing a child with special needs."

As they got closer to adopting a child, Myka became active in Facebook adoption groups where she asked other members to subscribe to her YouTube channel and interact with her videos. The way Myka describes it, she looked through hundreds of photos of children who were up for adoption before coming across the boy she and James would later adopt. The agency the Stauffers were working with told them that the child whose photo had so touched Myka might have a brain tumor or a cyst. "Without a doubt in our minds, we knew, no matter what state he came to us, that we would love him," she said in a video. "If anything, my child is not returnable." In October of 2017, the Stauffer family went to China to meet their adopted two-and-half-year-old son—whom they named Huxley—and brought him home with them. A now-deleted video of the moment had more than 5.5 million views on YouTube, and when they returned home and settled into life with Huxley, he became a focal point of their content.

In the story from *The Cut*, "Why Did These YouTubers Give Away Their Son?", Moscatello writes: "Myka and James asked viewers to invest not only time in Huxley's adoption story but money. In one video, they held a fundraiser asking their audience to donate $5 for puzzle pieces, that, when put together, would reveal a photo of Huxley." The puzzle piece donations were an early version of the face reveal Instagram post or YouTube video, in which a family vlogger shows their child's face for the first time, often racking up tons of views and shares.

The Stauffers shared lighter moments, like Huxley's first Christmas, alongside more serious content, like videos detailing his medical prognosis and communication challenges. And the content paid off: The Stauffers became more popular, garnering lucrative brand deals and millions of views on their videos. They shared Huxley's diagnosis of autism spectrum disorder level 3 ("the most severe form, in which communication challenges can be lifelong," Moscatello writes) and Huxley's therapist's assessment that he had an IQ below average and would likely never go to college. The Stauffers told viewers that it was a "struggle to vlog" because Huxley's behavior was so disruptive. Though they continued uploading videos into 2020, Huxley was featured less prominently—and viewers began to wonder why, pouring into the comment sections of the Stauffer's social media accounts to ask where Huxley was and what was going on with him. There were even anonymous accounts created specifically to ask questions and push for answers about Huxley's well-being.

In May of 2020, two and a half years after they posted a YouTube video documenting their family's journey to China to meet Huxley, the Stauffers made a video sharing the latest update on their adopted son: that, over a period of months, he had been transitioned to a new family and was happy with "his new mommy [who] has medical-professional training and is a very good fit." In an Instagram comment, Myka said she was "so naive" about the adoption process. "I was not selective or fully equipped or prepared. I received one day of watching at home online video training." The outrage was intense: detractors accused the

Stauffers of adopting Huxley for content and getting rid of him when he showed difficulties. And despite the fact that he was no longer their adopted son, their content of him was still featured on their monetized YouTube channel. He was still making them money, even though they weren't taking care of him. Soon after their video explaining the change, Myka deleted content featuring Huxley from her social media channels and took their family channel, The Stauffer Life, offline.

Myka hasn't posted to her Instagram account since June 24, 2020, when she posted a *Notes* app apology stating, "I could never have anticipated the incidents which occurred on a private level to ever have happened, and I was trying my best to navigate the hardest thing I have ever been through." Her husband, James, still posts to a YouTube channel called Stauffer Garage, which has 1.3 million subscribers and follows his car detailing, though his videos don't perform particularly well, averaging just tens of thousands of views compared to the average millions of views the family's vlogs once drew. No one in the comment sections of Stauffer Garage seem to know that the creator behind it is the family vlogger dad who became infamous for rehoming his adopted son. I read through hundreds of comments, all of which focused solely on the cars featured in his videos.

The Stauffers weren't able to bounce back from their scandal in a meaningful way. There's a clear line between what we can forgive and what we can't. A misstep is one thing; what some viewers saw as the neglect of a child is another.

Consider Piper Rockelle, for instance. Piper first went viral in 2016 when her mother posted a YouTube video of nine-year-old Piper making slime; in short order, she became one of the most successful child influencers in the US. She and her mother, Tiffany Smith, moved from their home state of Georgia to Los Angeles, to see what they could make of Piper's career. In 2017, Piper, who by this point had millions of YouTube subscribers, joined up with a group of other kid influencers to form a collaboration group called Piper Rockelle and the Squad, or "the Squad" for short. The Squad reached the heights of viral fame, reportedly bringing in hundreds of thousands of dollars *a month*. Their content included prank videos like the "Randomly Crying Through the Day Prank!!," viewed 3.1 million times, and challenge videos, like "Last to Leave the Trampoline Wins $10,000," viewed more than 4.7 million times.

But what seemed like a dream rapidly shifted into something more sinister in 2022 when eleven former Squad members sued Piper Rockelle, Inc., Tiffany Smith, and Smith's boyfriend and collaborator Hunter Hill, alleging "emotional, verbal, physical, and at times sexual abuse." The claims in the lawsuit are harrowing, like the allegation that Smith would assume the alter-ego identity of a character she called "Lenny the Dead Cat" and, speaking as Lenny, say sexually explicit things to the children in the Squad. Smith was accused of selling Piper's underwear to fans because, she allegedly told another Squad member, "old men like to smell this stuff." Smith and Hill allegedly did not compensate the Squad members for their appearances in the wildly successful YouTube videos, and they stand accused of sabotaging the ex-Squad members' YouTube channels after they left the Squad (supposedly through the use of bots and embedding the minors' videos into porn sites to get them blacklisted on YouTube). Smith and Hill denied the allegations against them. Though the defendants sought $22 million in damages, the case was eventually settled in 2024, with no admission of wrongdoing, for a reported $1.85 million.

The allegations against Tiffany Smith and Hunter Hill sparked serious conversations around how something so dark could possibly be happening behind the scenes of what seemed like normal adolescent YouTube content. Though the case against Smith, Hill, and Piper Rockelle, Inc. ended in a settlement, that wasn't the end of the story. In April of 2025, Netflix released a three-part docuseries called *Bad*

Influence: The Dark Side of Kidfluencing that explored the experiences of former Squad members, including claims that Smith would make the kids work ten-hour days and allegations of sexual assault by Smith. *Bad Influence* rose to the top of the charts on Netflix and became a viral topic of conversation across social media platforms. Though Piper and her mother weren't interviewed for the documentary (despite being asked "many times" to participate, a documentary producer tells me), they're the main characters of what is painted as a terrible story of parental greed, exploitation, and alleged abuse in the name of child influencer stardom.

In the aftermath of the documentary, I interviewed seventeen-year-old Piper for *Rolling Stone*. She told me of her child stardom: "I wish I honestly would have enjoyed what I had more because I would do anything to go back to those days." She said she never felt exploited by her mother but does feel she was exploited by Netflix. On the subject of the allegations against her mother, Piper said of her former friends, "They're not making it up, but they're extending the truth. I was there. I witnessed every day." Though I didn't speak to Smith for my *Rolling Stone* story on Piper, she did send me written answers to my questions, including the statement that "There's never been any kind of abuse. No physical abuse, no sexual abuse. No verbal abuse. Nothing."

But the allegations against her mother and the blockbuster documentary painting her childhood as ruined by the pressures of childhood fame aren't pushing Piper off the internet. If anything, she's extending her reach—she posts regularly not only on Instagram, TikTok, and YouTube but on BrandArmy, a website that draws comparison to OnlyFans. BrandArmy doesn't allow nudity, so creators don't have to be eighteen or older to sign up, but it is a subscription-based service where viewers can pay money for content that isn't available elsewhere; Piper's BrandArmy profile even says she takes requests. A few months after she turned eighteen, Piper joined OnlyFans and made $2 million in one day.

Perhaps the most notorious family vlogging scandal of all time came to a head in 2023 when former YouTube family vlogging royalty Ruby Franke was arrested on charges of child abuse. Prior to the arrest, Franke was the cocreator, with her husband Kevin, of the YouTube channel 8 Passengers, which had millions of subscribers who tuned in to watch the daily routines of the Utah mom and her six children.

In a typical video, Ruby would film herself getting the kids' lunches ready for school; more viral videos featured moments like her teenage daughter shaving her legs for the first time or her teenage son's voice cracking with the effects of puberty. In the Hulu documentary *Devil in the Family: The Fall of Ruby Franke*, Ruby's estranged ex-husband Kevin Franke says Ruby was the force behind the family's successful YouTube venture. "She filmed everything," he said. "When I came in the door, wherever Ruby was, the camera was." Just a few years after they began posting on YouTube, they had racked up millions of subscribers. The YouTube channel became a bona fide business—8 Passengers, LLC—which brought in $100,000 a month. Though there were moments that made some viewers side-eye Ruby—like the video in which she described refusing to bring her five-year-old daughter lunch at school after she forgot it—the Frankes seemed to be riding high as some of the most successful family vloggers in the game.

Behind the scenes, Ruby was introduced to Jodi Hildebrandt, a sort of life coach–therapist who would later be arrested for child abuse alongside Ruby. Under Hildebrandt's instructions, Ruby institutes strict consequences for her teenage son, Chad, including barring him from participating in sports and sending him to a wilderness camp. As a form of punishment, Chad was forced to choose between sleeping in the living room and sleeping on a bean bag. He chose the bean bag, where he slept for seven months. In a YouTube video, Ruby talks about the punishment. "When this video hit, all hell broke loose," Kevin said in *Devil in the Family*. The 8 Passengers YouTube channel lost tons of subscribers and the majority of their revenue. Instead of apologizing, as a lawyer proposed she do, Ruby doubled down and shifted the family's content from normal family vlogging content to religious teachings influenced by Hildebrandt, who by that point was doing counseling sessions with at least four members of the family. In short order, both Kevin and Ruby's eldest daughter, Shari, were excommunicated from the family, and Hildebrandt was installed as a sort of parental figure replacing Kevin. Though there were rumors floating around the internet about Ruby's descent into what Shari called Hildebrandt's cult, the situation came to a shocking head when Ruby was charged with multiple accounts of aggravated child abuse alongside Hildebrandt. The arrests came after Franke's twelve-year-old son appeared at a neighbor's house looking for help. Franke's son asked the neighbor for food and water, according to arrest records. The Santa Clara–Ivins Public Safety Department said the child appeared "emaciated and malnourished" and was seen with "deep lacerations from being tied up with a rope." Franke's ten-year-old daughter was found in Hildebrandt's home in a similar state. Franke went on to plead "with my deepest regret and sorrow for my family and my children," guilty to four counts of aggravated child abuse. She was sentenced to four one-to-fifteen-year terms in prison as mandated by Utah sentencing guidelines, meaning she'll be eligible for parole in as few as fourteen years but could potentially be imprisoned for up to sixty. Franke's business partner, Jodi Hildebrandt, was given the same sentence.

The story was an immediate media sensation. It was gruesome, dark, and almost too on-the-nose: the Utah mommy blogger who was actually concealing child abuse behind her viral videos. *This*, detractors claimed, was the trouble with family vlogs: You could never really know what was going on behind the scenes. In comments and videos and op-eds, people wondered: *Who would the next Ruby Franke be? What other family vloggers and mom influencers could be hiding horrors behind their perfectly edited videos?* The voyeuristic interest the entire internet seemed to take in the Ruby Franke case—and who, God forbid, might be next to follow in her footsteps—reveals something about the way we watch these family vloggers. We tune in because they're perfect and when it turns out they're not perfect, we're even more fascinated. What is more interesting than a perfectly glossy exterior? Apparently, the darkness it's hiding.

There was a time when I thought that, between the Stauffer and Franke and Tiffany Smith scandals, family vloggers and child influencers would fall out of favor—but that doesn't seem to be happening. Sure, the scandals are continuing to pile up, but they're not having an effect on the bottom line of the industry, which experts say is only growing. The top influencers and families are still making millions of dollars a year. Viewers aren't letting a few rotten apples spoil the bunch. If anything, there seems to be a quiet, desperate part of us that makes us *more* interested in the underbelly of this industry after we've gotten a glimpse of it. I've read so many Reddit comments wondering which other family vloggers abuse their kids like Ruby Franke did, or Tiffany Smith was accused of doing, and there's an almost gleeful undertone to the speculation. Underneath the clickbait headlines, documentaries, and exposés lies the question: Who is the next family that will fall? And when they do, what will we swear we knew all along?

The family vlogging industry and the advertisers and creators that populate it are largely untouched by the scandals that befall it. Kat Tenbarge, the internet culture journalist, said she's noticed that the truly terrible scandals of family vlogging like the Franke, Stauffer, and Tiffany Smith situations are treated as outside of the norm by both family vloggers themselves and the viewers whose attention fuels their careers.

"Both the blogging community and the advertising community . . . tend to view these types of situations as anomalies and outliers, and they're very careful to characterize them as such," Tenbarge says. "Even though Ruby Franke and Myka Stauffer existed within these communities for many years and their behavior was visible and it mirrored the culture within the community at large, as soon as it became clear that it would no longer be acceptable, people then flipped and started referring to them as these abnormalities, as these outliers. We know that's not true. We know that the content that they were making on YouTube is following along with the same incentivization structure as every other family channel."

The Frankes, the Stauffers, and Tiffany Smith didn't exist in a vacuum. They were incentivized, like every other family vlogger, to share increasingly personal details of their lives to feed the beast of the algorithm. But when that went wrong? Suddenly, they were held apart as examples of something darker instead of what they really were: a natural consequence of the commodification of a family's personal life. "The wider community, their peers in the industry chose to overlook these things, perhaps because it would require looking in the mirror and reconsidering your own incentivization and your own behaviors," Tenbarge says. "As soon as it crosses the line, then they're shunned and no one wants to talk about that, except to say how horrible it is. But it lacks a greater evaluation of the types of financial and engagement incentives that drove some of this behavior."

Despite the backlash from these scandals and missteps, the industry has flourished. "The industry has only grown," Tenbarge says. "It has never shrunk. There are only ever more people who are starting their own channels. There are only ever more people who are watching this content. There's only ever been more money flowing into the production and viewing of this content."

It seems that the scandals of the family vlogging world don't matter. There are the extreme cases—the Stauffers, which many viewed as child neglect, and the Frankes, in which there were actual charges of child abuse—from which the families can't rebound. But mostly? Viewers seem to have convinced themselves that the Frankes and the Stauffers and Tiffany Smith are strange deviations from the norm. And with that convincing, viewers are given the permission to keep watching, even as they keep an eye out for the next family who will fall.

As I write this, TikTokers are convinced that it will be Nurse Hannah, @hannah_bhiatt, who has around 580,000 TikTok followers. Hannah first gained her following as being the mom behind the viral "17 diapers" trend. If you weren't privy to that moment in internet history, the gist is this: One month postpartum, Hannah posted a TikTok in which she divulged that, while cleaning her house, she found seventeen dirty diapers lying around. "It's embarrassing, I know," she says in the video that went on to be viewed 19.3 million times.

The internet was divided: though some people thought having seventeen dirty diapers lying around your house was disgusting and weird, others (especially fellow newly postpartum moms) latched on to the video and started the #17diapers trend. Under the hashtag, moms shared their own versions of the seventeen diapers—the true, messy realities of being postpartum. The trend was heartwarming in a way the internet usually isn't anymore. As I was scrolling through the videos, I felt less alone as a new mom than I had in a long time. I even made my own video, sharing my struggle with finding time to feed myself during the day (and admitting that my sister had taken to feeding me spoonfuls of peanut butter to make up for it).

On an app full of shiny mom influencers and perfect family vloggers, it felt nice to share the truth of postpartum life. Nurse Hannah told *People* that the trend meant a lot to her and she was surprised by how "honest and raw" everyone was about their "deepest struggles as new and postpartum mothers." It was a nice story. Right? Even as #17diapers took over #momtok, there were detractors: people who thought Hannah was neglectful or a bad mom or just gross for leaving dirty diapers around her house. There were also accusations of rage-baiting: that she had staged the dirty diapers specifically to go viral and get a ton of followers.

We'll never know what prompted Hannah to post that video—whether it was a moment of true postpartum overwhelm or an attempt to go viral. Regardless, Hannah kept posting to her new followers—until she posted a now-deleted video that changed everything about her internet arc. In the video, her toddler son is sitting in a shopping cart and Hannah's husband, Braxton, approaches with something in his hands. When Braxton steps into the camera's eye, Hannah's young son rears back, in what people on TikTok were sure was a flinch evident of abuse. People downloaded the video and reuploaded it, slowing it down and zooming in to the young boy's reaction. "HERE" they wrote over the video with arrows pointing at the alleged flinch. There was this almost gleeful piling-on, with people quick to share that they had known since #17diapers that Nurse Hannah was a bad mom, and that this was only the latest evidence.

All over TikTok, moms took videos showing their own children's reactions to something being held in front of their faces. "LOOK," they said into the camera as they waved objects in front of their kid's faces. In the videos, their children laughed or looked confused or simply blinked but the message was clear: a child who had not been mistreated would not rear back in what the internet was now calling "the flinch."

The internet is often eager to take down its Character of the Week (double when it's a woman and triple when it's a mom), but the insistence with which people were claiming child abuse—and their willingness to use their own children to prove it—was chilling. Hannah was compared to Ruby Franke. In a response video, Hannah called the comparison "absolutely comical. We live in a world nowadays where people will turn nothing into something." She said her children were happy, healthy, and well nourished. In a separate video that now seems to have been deleted, Hannah said, "nothing is going on" and that her husband and son scare each other all the time, which is why her son reared back from her husband in the now-infamous video.

But her responses did nothing to quell the backlash. People found her home address and posted it online. They made videos detailing their calls to child protective services on Hannah's child's behalf and posted "deep dives" into her content, mining it for anything that supported their hypothesis. Though people are going viral with videos alleging CPS involvement, arrests, and investigations, any claims that Hannah has been arrested by CPS are unsubstantiated at the time of this writing. The Ogden Police Department has confirmed to ABC News that there is an ongoing investigation after "multiple reports" were submitted to child protective services. Neither Hannah nor her husband, Braxton, responded to my request for comment.

There's a part of us as viewers that seems to relish when family vloggers and mom influencers crash and burn—regardless of the children they take with them. I wonder if it's because we resent the influencers for having made it. They started as normal people just like us; then they entered the viral lottery and won, and their lives changed (seemingly, for the better). They move into nicer houses and buy luxury cars off the backs of our views. We're drawn to them because of their purported perfection, but we wait with bated breath for the dam to break. The only thing more interesting than perfection is a look behind it.

5

Filters, Feeds, and First Steps:

The TikTok Teen Mom Hustle

Every day, Brooke Morton (@brookexaloura) posts videos that give a window into her life as an eighteen-year-old mother of two. In one video, she lip-synchs to the popular audio "you might think I'm crazy" drawn from the opening bars of Ariana Grande's "34+35." The overlaid text reads "I got pregnant at 14, moved out of my parents at 16, had my 2nd at 17, just got engaged at 18, and plan on having my last baby in 3 years." In her TikTok bio is an email for her talent agency and a link to the clothing boutique she just opened. In her videos, Brooke is blond and bubbly and personable, talking straight to the camera about what it was like to get pregnant as a teenager. She holds her daughters, Aloura and Athena, makes them breakfast, or does her makeup. It's a day in a normal life, except for the fact that she's a teen mom.

On TikTok, the top teen mom creators boast platforms of millions of followers and make tens of thousands of dollars per month. Brooke has 2.2 million followers. There's sixteen-year-old Kylie, @the.noelanik, with 2.7 million followers. Her bio reads "Aaliyah's mommy <3" and has a link to the family page she shares with her boyfriend, which has 1.1 million followers. In the bio of that page, @theafam, there's a link to an Amazon page with her product recommendations, from which she earns commission. In Kylie's first ever video, she looks almost unbearably young as she stands in front of a mirror smoothing hair away from her face, wearing a bandeau with her bare belly protruding past her sweatpants. Braces cover her teeth. She is fifteen years old. The text over the video reads "Me remembering this isn't just a silly little pregnancy, but a human I'm now responsible for."

MariClare MacLamroc, who got pregnant when she was fifteen, has 2 million followers, and her bio includes both an Amazon storefront from which she earns commission and an Amazon wish list, from which fans can purchase things for her. In MariClare's most popular videos, she makes dinner in front of the camera, peeling frozen onion rings apart from each other and tossing them onto a baking sheet or breaking up beef in a frying pan. She has long, ornate nails, and half her hair is blond while the other half is brown. Her videos are hashtagged with phrases like #teenmom, #baby, and #15andpregnant, and she tells me she makes between $10,000 and $30,000 a month. Though that's not to say every teen mom creator makes as much as MariClare. There are smaller creators like Sarah (a pseudonym), who has around 320,000 TikTok followers or Nora (a pseudonym), with more than 159,000, who are both trying to gain a foothold in the world of teen mom creators.

The ethical considerations of what it means to be a parent influencer and feature your child in monetized content become even more fraught in the teen mom world, where many of the parents are still children themselves. As Kelsey Weekman, an internet culture reporter at Yahoo News puts it, "These parents are at the intersection of being a child influencer themselves and being someone who's *managing* a child influencer." It makes me wonder if teen mom content creators should be held to the same ethical and moral standards as their adult counterparts, especially when teen parents often face such intense financial difficulties and fraught living situations. Research shows that teen parents are less likely to graduate high school and more likely to make use of income assistance programs. Teen parents continue to experience economic disparities well into adulthood. With these grim statistics in mind, I can understand a teen parent choosing to make money on TikTok—even if they have to trade some of their and their child's privacy to do so.

Society's fascination with teen moms and the lives they lead is nothing new. MTV first aired their groundbreaking series *16 and Pregnant* in 2009, and the show became must-see TV. My friends and I were sixteen ourselves, and we became obsessed—we couldn't look away from the show's exploration of the lives of pregnant teenagers. They were both so familiar to us—the way they stressed over homework or fought with their boyfriends—and so foreign, as they navigated pregnancy and motherhood. *16 and Pregnant* ran for six seasons and went on to spawn the spinoffs *Teen Mom*, *Teen Mom 2*, *Teen Mom 3*, and *Teen Mom OG*. In 2025, *Teen Mom: The Next Chapter* was still airing, with Leah Shirley, Bentley Edwards, and Aliannah and Aleeah Simms still being followed as they turn sixteen, around the age their mothers were when they birthed them in front of MTV's cameras.

Leah Messer was first featured on *16 and Pregnant* as a seventeen-year-old expecting twins with her boyfriend, Corey. Looking back now, Leah looks so heartbreakingly young in that episode. Her blond hair was scrunched except for her straightened bangs, the going trend in 2009, and her West Virginia drawl was thick. As a teen mom born to a teen mom born to a teen mom, Leah's mother was only thirty-three when Leah gave birth at seventeen. Her grandmother was fifty. When I tell Leah about the teen moms of TikTok, she sounds older than her thirty-one years.

"Can I just be real with you for one second? Why are we still here? Why are we still here? From the bottom of my heart, I was hoping that *Teen Mom* was the epitome of, 'This is it. Hopefully, no other young girl has to go through this again,'" she says. "I still stand on that. To see [teen pregnancy] continuing to happen is very sad or it even being highlighted like it's cool. It's not—it's scary." What scares her about it?

"Young girls not being able to pursue their dreams and their careers. Watch me get emotional," she says as her voice breaks. "You don't realize everything that goes into being a mother at sixteen, seventeen."

During her fifteen years on television and in interviews, Leah has expressed so many regrets about following her mother and her grandmother into teen parenthood. But, she told me, of those repeating generational patterns, "it was normal. That was the life for many women before me. This was what was expected." Now in her thirties, she's doing her best to break the cycle for her own daughters, the oldest of whom are teenagers themselves. But I wonder if Leah can figure out why we're so fascinated by the story of teen moms. Why do they capture our attention so fully? "Maybe it's the struggle it creates," she says. "I don't know. It's sickening to me, though. It's disheartening."

Leah's time on TV has been marked by enough struggles to break even a casual viewer's heart. Raised in poverty, teenage Leah brought her premature twins, Aleeah and Aliannah, home to a trailer. She was in an on-and-off relationship with their dad, Corey, and by the time the twins were around a year old, it was clear there was something physically wrong with Aliannah. There are *Teen Mom* scenes that are still fresh in my head as Leah tries to fight through the medical bureaucracy to get her daughter's diagnosis, which ends up being an incredibly rare form of muscular dystrophy. Corey and Leah get married and then quickly divorced, and by the time she's twenty, Leah is married for a second time and expecting another child. While she was giving birth to her third daughter, Adalynn, Leah says she sustained a spinal-cord puncture and was sent home from the hospital with prescriptions for three different opioids. (It was 2013 in West Virginia, the epicenter of the opioid epidemic. The state would later reach a $400 million settlement with two drug distributors after alleging they had "recklessly oversupplied West Virginia with prescription pain medication," as reported by Reuters.) *Teen Mom 2* featured footage of Leah nodding out while obviously high and struggling to take care of her three young children as she falls deeper into drug addiction. She fights with her second husband, and they divorce too. The cameras watch as Leah struggles to gain a foothold in sobriety and create a life for her three daughters. In the latest season, Leah's twins are starting high school and, as the advertisement for the season promises, the story comes full circle, with Leah, who started on MTV as a pregnant high schooler now parenting high schoolers.

I know it's different for the teen moms of TikTok than it was for the *Teen Mom* girls—the TikTok girls are their own producers and they decide what goes online whereas the *Teen Mom* girls were at the mercy of (seemingly sometimes sadistic) reality TV producers. But I wonder if there's something to be learned from the experiences of the original *Teen Mom* girls. Is there anything Leah regrets about sharing her journey into and through teen parenthood? And can that serve as a warning sign flashing for the TikTok teen moms?

When I ask Leah if she regrets sharing anything specific, she doesn't mention the footage of her nodding off on drugs or fighting with either of her ex-husbands, but she immediately speaks of her kids. She knows that sharing Ali's muscular dystrophy diagnosis brought awareness to a little-known disease, but if she could go back in time, she might share less. There are some moments that she'd take back to protect her daughters' privacy, like filming their toddler tantrums or baths. Watching her own teenage parenting from the vantage of her thirties, Leah sometimes cringes at how she reacted to her children, with repeated scenes of her losing her temper or getting frustrated. "I would definitely parent differently at thirty than I did at seventeen," she says. "But those things did get documented."

And that documentation changed her life, Leah says. Though she has her real estate license, she hasn't been able to get a job, which she thinks is partially because the companies she's interviewed with don't want to be known for having the *Teen Mom* girl on staff. "It's harder to even get a stable job in your community," she says. "Now that you've documented that in the media, that's forever out there. It's almost like you're having to fight to regain the trust of everyone around you because they don't really know you. They still know you as that seventeen-year-old. It's like you're a hamster. It's like you can't get ahead. Let's say, you have your five minutes of spotlight on TikTok or on TV. Then what? You know what I mean? Where do you go from there?" If she could tell the TikTok teen mom girls anything, it would be this: "Be mindful of what you put out there for the world. Not everyone needs to know everything. Some privacy is okay."

It can be hard to convince teenagers that they may grow to regret their TikTok fame. "That's something that in terms of media literacy, young people really don't seem to care about until they're suffering from the consequences of it," says Kathryn Jezer-Morton, a columnist at *The Cut* who holds a PhD in sociology. "That fame—or relative fame—has a real cost." The way I see it, the cost is multifold: There's the privacy you're trading away for TikTok views; then there's the brand of Teen Mom that you're stamping on your forehead. Like Leah said, that label can follow you throughout your life, even into your midthirties. But from an ethical perspective, Jezer-Morton sees the question of whether teen moms should be held to the same standards as adults as cut-and-dried: "Kids of a teen parent should have no less access to privacy than kids of an adult parent," she says.

I get that. I really do. But the allure of an influencer paycheck is difficult to ignore. What other job could MariClare be doing that would net her over six figures a year as a teen mom who only received her GED? How else would Kylie have been able to buy herself a car at sixteen? What would allow Brooke to be shopping for houses as an eighteen-year-old mom of two?

"This whole thing is very dystopian," Jezer-Morton says. "If we're seriously saying that influencing is the only way that a good-looking teen mom can expect to support herself, we are so fucked. It is so bad." It's not that she can't see the justifications—daycare is prohibitively expensive and teenagers are generally only qualified for pretty low-paying jobs—but our economic system has failed if some of the most vulnerable people among us (i.e., teen parents) have to enter the viral lottery in hopes of being able to take care of their families.

When you watch the teen mom royalty of TikTok, being a teen mom doesn't seem like that bad a deal. In their videos, they snuggle their cute babies and buy cars and start businesses. The content doesn't carry the same desperation as those early days of *Teen Mom*. It looks almost cozy—maybe not a lifestyle to strive for, but not one that is dripping in desperation or difficulty. For girls who find themselves pregnant as teenagers, TikTok can be a way to manage the inherent difficulties of being a teen parent.

When I first spoke to Nora, she was sixteen years old and thirty weeks pregnant. Nora found out she was pregnant when she was already nineteen weeks—she was basically symptom-less, except for having to pee all the time. When her mom and a friend wondered if she was pregnant, Nora took a test that immediately turned positive. "I remember I called my mom and I'm screaming and crying on the phone. I was going crazy. And honestly, a thought dipped in my head like, *I can't do this. Like, I have to figure out another option because I can't have a baby*. But as soon as I woke up the next morning, I was just like, *Absolutely not. Like, I'm keeping this baby one hundred percent*." Before getting pregnant, Nora planned to graduate high school early and head to college out of state. "Obviously that's not happening now," she tells me. "But also I never wanted to have kids either. I wanted to just be single, no kids, nothing. Just work. Yeah. So it changes a lot."

Before she was pregnant, Nora watched teen mom content. She was a follower of Brooke Morton and watched TLC's teen mom reality show *Unexpected*, but she never expected to find herself in the same situation. Sometime in her fifth month of pregnancy, Nora decided to start a TikTok documenting her journey. She figured it would be cool if the account blew up like she'd seen happen for other teen moms. Maybe that could happen for her too. In her first TikTok payout, she made $560 in one month. "That's like, good money for just posting," she said. "I think it would be so cool if I could expand my TikTok and then even [further] to other forms of social media and make good money off it." What if her page takes off like Brooke's or Kylie's has? Nora gave birth to her son, but her TikTok growth has slowed. Though she had early success, gaining tens of thousands of followers in the first month that she created her account, she still only has 159,000, far from the millions that the biggest teen mom creators command. She does aspire to get bigger and make the kind of money they're making, all in the hopes of being able to support her son by herself.

I'll just get ahead of what you might be thinking: I don't believe that Nora or any of the other teen mom TikTokers got pregnant on purpose to become content creators in this lucrative niche. I suppose it's possible, but I think it's much more likely that each of them was just an unthinking teenager who made a misstep that led them to pregnancy before they're even legally allowed to vote. And once they found themselves in that position, young and pregnant and with few options for taking care of their future children, they started thinking. They remembered the success of the *Teen Mom* girls or the *Unexpected* girls or the TikTok girls, and they opened the app and started an account.

Each of the teen mom content creators I talked to insisted that the money was a bonus, but that their main motivation in creating content around teen parenthood was to give other girls in similar positions assurance that they weren't alone and they could manage the monumental challenge of teen motherhood. Take Sarah, who was fifteen years old and fourteen weeks pregnant when I spoke to her. She found out she was pregnant after going to the hospital for stomach issues. When the medical staff asked if she was sexually active, she said yes and they gave her a pregnancy test, which came back positive. Her mother, who was also pregnant, was with her when she found out. "I was terrified that I wouldn't be able to provide for my baby," Sarah said. She was sure she didn't want an abortion—she "doesn't care" if other people have them, it just wasn't for her—but her fear was overwhelming. "I was just terrified that I wouldn't be able to provide for the baby."

Sarah's TikTok account has since grown to over 320,000 followers. Her first post was a series of photos of foods she was craving (including cake, Taco Bell, and Caesar salad) and was hashtagged with #teenparents. Sarah said she wanted other teen moms to know they could handle what was ahead of them, though she's aware of the narrative that she and other teen mom creators are glamorizing teen pregnancy. "I think [people] think that because in my videos, I'm very happy and I don't sit in my room and cry about the situation all the time, they just think, 'Oh, you're normalizing it. You want this to happen.' But in reality, no, this is not what I wanted for myself. I wanted my career, and I'm still going to get to my career, but this is a little step in the way."

Sarah does look happy in her videos, as she gets ready for her gender reveal party and goes shopping and hangs out with her boyfriend, all while her round belly grows with each post. I can understand the concern that this content is glamorizing or normalizing teen pregnancy (though the teen birth rate in the US has been falling since 1991 and is now at a record low, according to data from the Centers for Disease Control and Prevention). It doesn't look so bad from the perspective of these TikTok videos, in which young moms gain fans and platforms and post collages of their ultrasound pictures. But I wonder what we want from these girls. Do we want them crying on camera and documenting their struggles to pay bills as they fight with their families and boyfriends? Do we want to see them being ostracized at school for being a pregnant teen or dropping out to go to online school? Why do we want to see them struggle? Is it because we think they deserve it?

There are some truly harrowing scenes in those early seasons of *16 and Pregnant* and *Teen Mom*, where parents of the teens cry at the thought of their babies having babies. The young couples scream, fight, break up, and leave each other on the side of the road. Newborn babies are brought into impoverished homes that were already struggling for resources. When my friends and I watched as teenagers ourselves, we were aghast at how difficult being a teen mom looked. The show really did seem to raise awareness about the challenges of teen pregnancy: One study found a 5.7 percent reduction in teen births in the eighteen months after *16 and Pregnant* first aired. Teen motherhood on these shows looked terrifying. Later, when the moms started earning six-figure paychecks from appearing on the reality show and were able to buy new cars and new houses and quit their jobs, they were accused of glamorizing teen pregnancy. But were they glamorizing it or just trying to survive it?

Sarah made $500 in the first two weeks her account was monetized ("which is not bad," she tells me, and I think about how it would have taken sixteen-year-old me nearly seventy hours of work at my job at Jamba Juice to make the same amount of money). When I ask Sarah if she looks at the wild success of other teen mom content creators and hopes to emulate it, she's unequivocal. "Yeah, I am [hoping that]," she says. "I feel like if I actually do blow up more and I do get more money, I feel like that would be great." But, she adds again, her goal isn't money: It's raising awareness about the reality of being a teen mom, even though her content is generally upbeat and gives a rosy view of the whole experience. And her mom is totally supportive of the endeavor, Sarah says. "She always tries to give me video ideas and stuff. She's happy that I'm earning money for [the baby]. She's very proud of me, actually, for stepping up and doing what I have to do."

Why is the public so obsessed with depictions of teen motherhood? From *Teen Mom* to the teen moms of TikTok, we seem to be unable to look away from pregnant and parenting teens. Kailyn Lowry, who first appeared on *16 and Pregnant* and later *Teen Mom 2* as a pregnant sixteen-year-old, compares society's interest in teen moms to our obsession with true crime. "It's because it's not normal, right? That's not how it should go. It's almost unfathomable."

There's another way to understand the interest we have in teen moms, and it's decidedly darker. This is how Rilah Ferrer, now a twenty-two-year-old mom who appeared on TLC's *Unexpected* after becoming a mom at sixteen, sees it. "I feel like people like to see others in a vulnerable situation," she tells me. "They get entertainment off of that, just seeing girls in a bad situation. I feel like that's people's entertainment, [like] 'These dumb little girls get knocked up.'" When Rilah responded to the casting call for *Unexpected*, it was because she was in a desperate situation as a pregnant teenager and thought the show might change her life. She needed money, and she was told that being on the show would open other doors for her. "The casting people were basically making it seem like if you sign up, you get a big break," she said. "The TV shows really want to push the narrative like, 'Oh it's to educate other girls about safe sex and teen pregnancy.' But no. I think it's really just to exploit girls in vulnerable situations for ratings and money. It's just for entertainment." Rilah tells me that being on the show didn't change her life in the way she expected. There was no big break, and now the details of her chaotic relationship with her daughter's father are documented forever. Even as *Unexpected* plays in reruns and is available on streaming sites, Rilah and the other girls featured in it don't get paid residuals. They didn't get paid at all beyond the original fee (that Rilah and other sources have said was only a few thousand dollars). Though she was asked to return for another season, Rilah declined the offer. She doesn't want the intimate details of her life on TV. "These people are asking me back because they want to exploit my dirty laundry. I remember thinking, *It's one thing to exploit myself being pregnant, but I'm not going to show my daughter and have her father shown out to the world like that,*" she said.

And here is the heart of the issue of teen mom content creators. These girls are kids themselves, under the age of eighteen, and they're in situations where they have to decide to trade their privacy—and the privacy of their children—for money.

Nora, the teen mom TikToker we discussed earlier who found out she was pregnant at nineteen weeks, seems to have given the question of privacy and boundaries a lot of thought. She doesn't feature her son's father in her videos, and she doesn't discuss the status of their relationship. In her most popular video, she says they aren't together but "things are good." No further details are given, even though I'm sure airing out the situation would have propelled her past her current 159,000 follower count. At first, Nora didn't show her son's face in her videos; in the post announcing his arrival, she covered his face with a blue heart emoji. "I try to keep certain things limited of the info I give," Nora says. "I don't want [my son's] pictures and stuff all over the internet." She describes another popular teen mom creator, who immediately posted photos of her newborn from the hospital; within hours, there were fan accounts posting the photos themselves. "I don't want my baby to have none of that," Nora says. "I don't know. I guess it's just like, that's your baby, you know? Like, I don't want him all over the internet for everyone to see. Not immediately at least." By the time Nora's son is four months old, when he appears in her videos, his face is showing. When I ask her about her change of heart, she texts me, "honestly he's just so cute I had to share" followed by another message: "I definitely still am limiting to what is seen of him on social media still," followed by one more: "I don't want him on my media too much."

When Sarah was still pregnant, she told me she wasn't sure if she was going to show her son when he's born. "I haven't really thought about it," she says. "I feel like at first, I probably won't show the baby. Maybe when he's a little bit older, I will. But I do know there's a lot of creeps and weird people. So I'm not sure. I'm going to talk to my boyfriend about it. It's not really something we've been talking about. So, yeah, I don't know. We'll see when the baby is here." Within a few days of giving birth, Sarah vlogged her son's birth and his first week in TikTok video after TikTok video. Sarah now has more than 320,000 followers and recently posted a vlog explaining her decision to quit her job at Dunkin' to focus on content creation. "I do know I am super blessed and privileged . . . to be able to just literally sit in front of a camera, talk and show my life, and be able to make income off of that," she says. "My Dunkin' era is over. I like my social media. I like my content creating. I've always wanted to do this since I was younger."

Meanwhile, in the comments of Brooke Morton's videos, TikTok users discuss how much her eldest daughter, Aloura, has grown. "Been here since the beginning!" they write. "So cute she's growing up too fast," another one reads. "Literally my favorite internet family!" says a third. (Can we just stop for a second and appreciate the phrase *my favorite internet family*?) While reading the comments on sixteen-year-old Brooke's pregnancy announcement for her second daughter, Athena, I'm struck by how invested people are in Brooke's life. "OMGG CONGRATS. aloura's gonna be the best big sis," reads one comment. Another: "OMG CONGRATS I LITERALLY SCREAMED IN PUBLIC." Brooke's popularity as one of the preeminent teen mom TikTok creators can be explained by a few things: for one, she's thin and blond with blue eyes, like many popular mom influencers. She has a striking story, even amongst teen moms: She got pregnant when she was only fourteen years old. She also has a really cute little family: Brooke and the father of her children, Jason, are still together and recently got engaged, with the engagement teased through a series of posts and then announced in a video that garnered 1.2 million views.

I don't think it can be ignored how present Brooke's children are in her videos. Aloura or Athena, or sometimes both of them, are featured in the majority of her posts. Even Brooke's handle, @brookexaloura is an homage to her first daughter. In her comment section, users fawn over Brooke's daughters: "I love your little family your girls are so beautiful," "Oh you have the cutest girls," "Your baby has grown up so much." Teen mom creators, in the end, are still mom influencers. Part of what draws people to mom influencers and teen mom creators seems to be the same thing: the kids.

The other part of what makes teen mom TikTokers so fascinating is something that's uncomfortable to admit. I think Leah and Rilah were right: I think viewers like to watch young people in vulnerable situations. We want to see their stumbles as they figure it out, or we want to see them fall entirely. There's a part of us that likes watching the content of someone we can feel so easily superior to, and who is it easier to feel superior to than teen mothers? But as I and millions of other people have seen, the top teen mom TikTokers have spun their struggles into gold. Let's put it this way: Kylie, after getting pregnant at fifteen, paid for her first car in cash, something I've never done at the age of thirty-two. So who in this scenario is really superior?

6

Curated Childhoods:

The Parents Behind the Posts

Sam (a pseudonym) knew their kids didn't want to be filmed. The brand deal was for a stupid product, and the kids didn't want to be involved, mostly because they would have to dance, and they hated dancing on camera. Sam understood that, but the company was going to pay the family $40,000, so everyone needed to get in line. "The kids were miserable," Sam remembers. "They hated it. They hated all of it. But still, it was like, we've got to do this. It would boil down to, 'Okay, what do you want? You do this video, I'll pay each of you $500. Right here. Right now. Here's the cash.' But even then, that wasn't enough. The kids were like, '$500 is not worth this.'"

What happened after the kids rejected Sam's $500 bribe? Well, Sam says with a sigh, the next tactic was to "manipulate and guilt trip the kids. 'This is our business. This is what we do. This is our family job. If we owned a farm, you would all be out milking cows. No questions asked. If we owned a restaurant, you'd be helping wait tables. No questions asked. This is what we do. So get in line and let's do this. And then let's go out for ice cream, and I will pay you your big fat wad of money.'" The most Sam ever bribed the kids to participate in a brand deal was $1,000 cash each for their participation in a 1- to 2-minute video.

Sam and their spouse started their family vlogging YouTube channel in the mid-2010s and grew it into a platform that boasted millions of followers and over a billion lifetime views. Their kids—whose ages spanned from around infancy to teenagers—spent much of their childhoods and adolescence on camera. When a scandal forced Sam and their family offline (giving more details than this would compromise their identity), Sam's older kids struggled to find their footing in the normal young adult world. Sam's oldest daughter told them they didn't need to get a job, that working was for other people. Sam's oldest son got a job in the construction industry but quit within a week, telling Sam he would make only $15 for an hour of work, could Sam believe it? He felt he was totally getting taken advantage of, and he didn't want to go to college either. Money had always come so easily; the kids couldn't conceive of a normal work situation. Sam's younger kids grappled with the change in circumstances too. Life on the other side of YouTube fame was not what they had become used to. "They really struggle when the attention is not on them. They really struggle when things don't go their way, or they don't get what they want, or they don't get bribed to do what other children are just expected to do."

Sam and their family spent nearly a decade living in the light of family vlogging fame, and during that time, the kids got used to a luxurious standard of living. When they went on vacation, they stayed in the best hotels. When they didn't want to do something, they were bribed with money. But then the family channel went offline; the money stopped pouring in, the luxury vacations ground to a halt. And now Sam worries that the kids have been irreparably ruined by the years they spent internet famous. "I spend all of my focus and time trying to prepare my children for a very normal adulthood life, normal healthy relationships, a mundane normal job, standard or an average standard of living, those kinds of things," Sam said. "Helping them transition to the idea that what you make of your life is directly correlated to the work that you do and the sacrifices you make and the level of effort you put in and that you can't just expect money to drop out of trees for doing ten minutes worth of work. That's not reality."

But money dropping out of trees had been the reality of their lives for so long—and they had learned the lesson well. *Was it too late?* Sam wondered. Had Sam and their spouse raised out-of-touch egomaniacs who would never understand the value of a dollar? Sam worried that the die had been cast. "Family vlogging is a narcissism factory. You are literally creating a narcissist. Why not? They come to believe that the whole world revolves around them, that they are the center of attention. They're what everybody wants to watch, see, and be friends with, and like." Sam pauses for a moment before continuing. "And you'd be a fool if you think that they just grow out of that, and they don't carry that with them into adulthood."

Sam tells me that at the heart of every family vlogging journey is insecurity. The parents want praise for their parenting skills, their beautiful children, and their lovely home—and they find that praise through online fame. Sam's spouse, whom we'll call Cameron, was the one who wanted to start family vlogging after seeing how lucrative it was becoming for other families in their community. At first, Sam hated the idea of becoming a vlogging family. "It felt like a violation of our family's privacy," they said. "It felt like the camera was violating our space. And there was a strong sense [of] 'Who the hell cares about us?' Like, you know, we're not special. We're not celebrities. We're just average people who live in an average community. And you know, my [spouse] would always say, 'Well, that's exactly why people are interested, because we're real, and people are curious and we have, you know, some light, some warmth, some encouragement, some hope to share with the world.'"

Sam didn't want to be seen as unsupportive—plus, they figured the foray into family vlogging was a lark that wouldn't last. When Sam's spouse stayed up late into the night editing videos, Sam didn't say anything. "It was like there was a drive, something pushing [Cameron]," Sam says. What was pushing them? I ask. "Determination to get recognition and praise," Sam says.

Before long, Sam realized Cameron wasn't going to give up, and they decided to get involved, too, encouraging Cameron to network with other creators in their community. The couple soon realized what many a family vlogger before them had: Content that featured their children did much better than content without it. "And so the videos began to focus more on the kids and more on the kids' lives," Sam says. Even better was if the children were in some type of vulnerable state, if they were emotional or something embarrassing was happening to them. Sam started paying attention to video analytics, treating the entire venture like a puzzle they could solve if they paid enough attention.

The family's first paycheck, which came a few months after they started their channel from YouTube Adsense—a revenue share program in which qualified creators are paid per views—was only a "couple hundred bucks," but it was enough to show them that there was potential in the venture. "That changed everything for me. It showed us that there was legitimate promise in this." Soon enough, the family's channel was "growing like crazy. Every month the money started to be more and more and more. And then pretty soon, like, your filter just goes right out the window."

What filter? I ask. "Well, that moral filter of like, 'Should we be filming this?' You don't even care about that anymore. It's just like, will this attract views? Yes or no? Yes. Okay, how do we present it in a way that's consistent with our brand, consistent with the messaging we're wanting to share, but still exploit it somehow? So for the sick child, it was like, okay, we're going to show the child being sick and vulnerable, but then we're going to show, like, what supportive and loving parents we are. That's what we were trying to portray, that like, we're loving parents. Praise us."

Did Sam feel praised? I ask. Yes, they tell me. The comments were flooded with people telling Sam and Cameron what wonderful parents they were, how lucky their children were, how the commenters wished they could be their children. How did that feel, to be so praised?

"Like crack," Sam says. "It was like a drug."

Once the family's channel really took off, they discovered that they could make more than Sam's entire yearly salary in just two or three brand deals. Everything was potential content, and soon nothing was off-limits: not a child vomiting or crying or being potty trained. I ask Sam what the process of filming was when they or Cameron realized something was happening that would translate to views. Sam tells me about a time that one of their children was sick and vomiting and Cameron filmed it.

"In a vlogging family, the camera becomes just another member of the family. The camera is always present. And so you just kind of become attuned and numb to it. So whatever happens in the house that day is fair game. And vloggers and momfluencers and such, they come to realize that nothing draws clicks, nothing draws views, nothing draws attention more than drama and vulnerability. That is—in terms of clicks and views and stuff like that—that's gold. And drama is gold too. And so if there's anything that is vulnerable, if there's anything that is dramatic, the camera's there. No permission is asked. No, you know, it's just there, and it's just filming."

A vomiting child, in the case of a vlogging family, is potential—potential content and views and likes and subscribers. They relish these moments, Sam tells me, when the unexpected happens and they can exploit it. I ask Sam what was going through their mind as Cameron filmed their vomiting child.

"Any parent influencer is constantly fighting feelings of guilt, feelings of questioning, second-guessing yourself. But you learn to bury it and silence it with justifications," Sam says. Justifications like: *We want to be as real and vulnerable with the world as possible*, and *It's courageous to show the reality of a family's life*, and *We're uplifting and encouraging others with our content*, and *We're helping people, we're giving a gift to the world.* "The gift is ourselves, our family," Sam says. "Every influencer family I've ever met, including ourselves, had this inherent belief that they had something to offer to the world."

Everyone in the equation is getting something in return. The vlogging family is making money hand over fist and receiving the adoration of the viewers. The viewers feel uplifted by the content or encouraged or comforted in some way. And the platform—well, the platform arguably does the best, because they're making tons of money from the advertisements threaded through the content.

It was only later that Sam wondered if the kids were the ones losing out.

As Sam's family got more popular in the influencing world, they were welcomed into rarefied circles made up of the most popular family vloggers. Those families had entire teams on hand to handle production and costumes and lighting and search engine optimization and scripts. They had marketing teams and agents who negotiated on their behalf. Once, Sam and Cameron were on vacation with several other vlogging families. The parents were sitting around hanging out together while the kids played by the pool. There was one family there who had particularly leaned into the tendency to post videos with thumbnails showing their children in shock or full of fear. "We were just sitting around talking and somebody asked him, like, 'Do you ever, like, get queasy about it or nervous? Like, do you ever have second thoughts?' And he laughed and he's like, 'Well, I learned to bury my conscience years ago.' He's like, 'Trust me. If I knew that a video would explode if me and my kids were dressed up in penguin costumes jumping off our roof, I would do that.' And everybody laughed. What he was basically saying is, I will prostitute myself and my children for my viewers if they will continue watching me. That's really what he was saying."

I ask Sam if they related to what the dad said—that he would do anything for views. Yes, Sam says. To a degree, they did relate.

"The misgivings come early on, and I think that's true for everyone," Sam says. "The deeper you get into it, the more you numb out from it. And I think this is probably true for every creator. The biggest fear that they have is losing the influence they've gained, so they'll do anything they can to keep it. There's no room for misgivings. There's no room for a conscience at that point."

And how could there be? They were making over a million dollars a year.

Heather Bell is the matriarch of the Bell family, known as @justthebells10 on TikTok (2.9 million followers), Instagram (792,000 followers), and YouTube (389,000 subscribers).

She first went viral in 2020 when she made a video about her son, Joshua. The video has since been deleted but subsequent videos detail Joshua's early life as affected by shaken baby syndrome. In one video that has been viewed 2.2 million times, Heather shares that when she and her husband adopted Joshua when he was one and a half years old, he couldn't walk, crawl, or feed himself. The video is tagged #myhero, along with #shakenbabysyndrome and #survivor. "People love Joshua," Heather tells me of that early virality. "Then people wanted to see all the kids. We have a blended family. We adopted seven children and two of my kids are biracial. People just wanted to know how we did things. And so that's really how things started taking off was just sharing how we manage things as a big family."

It's easy to see why their content took off during the lockdowns of the pandemic, when so many of us were stuck inside our small apartments, dreaming of the outdoors, and the Bells were a family living in the country on a huge farm. Plus, families like the Bells tend to do well on YouTube. In her thesis "Anything for Views Parenting: Framing Privacy, Ethics, and Norms for Children of Influencers on YouTube," ethicist Bridie Hamilton wrote, "Large families, religious families, blended families, adopted children or children with disabilities, and families living in exceptional locations or under novel circumstances have found niches on YouTube."

For Heather, the sudden popularity was intoxicating. She shared and shared, answering any questions people left on her videos and carrying a camera with her everywhere she went. Those are things she regrets now. Looking back, she wishes she had never posted a house tour. If she could do it again, she'd share fewer details of her kids' adoptions.

"I shared more than I should have shared," she says. "I guess I was just so caught up in giving people what they want, right? They want to see this, you give it to them. And [I was] just so obsessed with having followers and likes and views. It just consumed me."

There were a few instances that changed her mind on what and how much she shared. First, she could tell her family was getting irritated ("They were just like, *Oh my goodness, really? The camera again?*"). Second, she began following Sarah Adams of @Mom.uncharted, a TikTok creator who uses her platform of more than 308,000 to advocate against sharenting. In one video, Sarah told family vloggers to ask themselves if people would still follow them if they didn't show their children. Would they still have a platform? "When she had said that, I thought, *Wow, I wonder if it was just me, would people still follow me?*" With that push, Heather fell into a rabbit hole of articles, TikTok videos, and Reddit posts of the supposed dangers of family vlogging, including privacy issues and child exploitation. She decided to "recalibrate."

While she's telling me this, she chokes up. "First, I sat with my family, and I apologized to my husband and to all my children because social media was consuming me. I first had to get it right with my family, and then I went on to my page and deleted anything that had to do with the house tour . . . and anything private about my kids' adoptions. My son Joshua, we adopted through foster care, and he's a shaken baby. I would share things about him. It wasn't detailed stuff, but it was just how I'm so happy that he's alive. But people didn't perceive it that way. I don't want to be a stumbling block to people, and I don't want them to think that I'm trying to be intrusive on my family. I had to go back and I had to delete a lot of stuff, even things that went viral big time, like millions [of views]." (Scrolling through Heather's content now, I don't see any house tour videos. But there are still videos describing her children's adoption journeys and Joshua's experience living with shaken baby syndrome.)

The change of heart left her thinking about what was next for her platform, and for her family. "I had to just go and completely regroup my page," she says. "It's been very freeing not feeling that I have to share these intimate things because no matter what any family or mom tells you, you still don't feel one hundred percent when you put up that stuff. You might act like you do. You might think it's, *Oh, we're just sharing this*. But deep inside, it's not right."

So why did she do it?

"For the likes. For the views. To grow. And honestly, social media would determine how I acted. If I didn't get a lot of views, I was crabby with my family. If I was viral, I was happy."

It was humbling to have to apologize to her children, she says. There are videos that stick in her mind: ones where the kids look irritated at their mother pointing a camera at them yet again or one that she posted about her then-teenage son Robert, in which she said he was looking for a girlfriend. "He was so mad at me," she remembers. There was another video in which Robert audibly says, "Of course, here comes the camera." What was it like to realize she had taken things too far?

"Well, I felt stupid and I felt humbled," she says. "It made me feel like, *Wow, get your act together, Heather*. Because I think as an influencer, I mean, to be honest with you, one of your first instincts is to capture the moment, right? Capture the moment, and then put it on social media. And so that's what you're always thinking, capture the moment, capture the moment, or making your kids redo that moment you didn't capture. 'Hey, can you do that again for the camera?' I did feel like . . . Gosh, I just ruined whatever we were doing. I ruined the whole vibe of it.'"

Now, her content is more focused on the family's lifestyle rather than the personal details of specific kids. She makes videos about how she cooks for her family of ten and the preparations for her daughter's wedding. "Are we having as many likes? Nope. Do we go viral as much? Nope. But I don't care because that's temporary. My family is forever. And I want my family to be around me. I think my kids are more at ease now."

There's a freedom in the way Heather approaches social media now. She doesn't feel as beholden to the content machine, the way it is constantly churning. She doesn't spend so much time filming and editing. Her house is cleaner. She's baking again (although sometimes that is filmed and edited into TikTok videos). She's crafting and canning and lying out in the garden under the sun for hours. And her neck doesn't hurt anymore from being hunched over her computer and phone. "It's so nice," she sighs.

The Bethune family, who went by the handle @beingbethunes and have 271,000 YouTube subscribers, started vlogging after a tragedy prompted the family to rethink their lives. In 2011, the family was on their way to Disney for their son Ethan's seventh birthday when they were in a car accident. Ethan was killed instantly. "Really that was the catalyst to us realizing that you only live once and life can change in a moment," Jenn, the Bethune family matriarch, tells me. "My husband said, 'What if we bought a vintage bus, traveled the country, and documented it?'"

The family had been traveling and creating content for a few years when another family's tragedy catapulted them into virality. When twenty-two-year-old Gabby Petito went missing and her case captivated the nation in 2021, the Bethune family posted a video that included footage of what seemed to be Petito's abandoned van. The video, which now has over 2.7 million views, was titled "Is This Gabby Petito's Van Caught on Youtuber's Camera? Read Description." This video is easily their most viral ever. The video with the second most views is also focused on the possible sighting of Petito's van. The third most popular video on their channel, which is a tour of their bus, has only 318,000 views.

After that brush with virality in 2021, the family doubled down on content creation, making videos that highlighted the reality of living with three children in a converted bus. Their videos were doing well, and they were gaining subscribers, but the popularity came with a price familiar to anyone who has ever had a piece of content take off online. "I had millions of people critiquing every aspect of who I was, how I talked, what I looked like," Jenn says. "They were telling me I was like Hitler and I should die [or that] my kids sleep in coffins. And I'm just sitting there like, wait, wait, wait. I'm just showing you what we do. Like, this is just our life. This is what works for us." Jenn came to believe that the people who were hating on her were triggered by her ability to break out of the mold society expects of parents and families. And throughout all the hate, she felt a call to keep creating.

At one point, Jenn says she and her husband stopped featuring the kids in videos "because we did not want virality or views through means of our children." But that didn't last. "The kids had started asking about being in the videos again," she says. So the family started having conversations about the content they were creating. Jenn would have an idea for a video and ask the kids if they wanted to be part of it. If they did, she says, that was great—and if they didn't, no problem. (Every vlogger and creator I talked to for this book told me a version of this. I can't help but wonder if it's true.) Her oldest son, Ben, doesn't like to be on camera so he's featured less in the family's content. Her daughter Molly loves everything about content creation. "She writes, directs, edits, does all of her own things. I just kind of neaten it up a little bit. She writes her own description, does her own hashtags," Jenn says. "She's not allowed on Instagram because I don't want her brain consuming that blue light and all of that, the stuff that social media has." Instead, Jenn posts the content for Molly and monitors it. "I'll go through the comments, and if there's great comments, I'll read them all to her, and sometimes I'll screenshot them so that she can read them."

In Jenn's explanation, I sense a motherly urge toward protection, but the logic behind her boundaries seems amorphous. On one hand, she allows Molly to create and post public content. On the other, she keeps her daughter from direct interaction with comments or messages. What is she trying to protect Molly from? "We have seen the darkest depths of the internet," Jenn says. "And she does get to go on technology. She gets to go on a phone and text her friends. She's got games that she plays, she plays Roblox. We just don't want them on social media."

I've heard this over and over from influencer parents—they're both using social media constantly themselves and featuring their children in their content while trying to limit their kids' own involvement in ways that sometimes seem arbitrary. Does it matter that these kids don't have access to their own Instagrams if they're still creating content that is posted on them? Where is the line exactly? And what exactly are parents trying to keep away?

I ask Jenn this as delicately as I can, and she tells me that she knows there are perverts online because she once received a "nasty, derogatory comment" about Molly on a video she was featured in when she was eight years old. "It was talking about, like, what he could see," Jenn says. "He wanted her to pull up her skirt so he could see her underwear." Jenn's retelling of the comment puts a pit in my stomach—it seems like the validation of people's worst fears of what it means to show kids online. What did Jenn do in response to the comment? "I immediately deleted and blocked and reported," she says. "That's all you can do."

I can't help but think: *Is that all you can do? Should she have taken Molly off social media entirely after that incident? Or would that be an overreaction, to limit this means of expression because of what is hopefully a rare occurrence?*

"When something comes up, like that comment about Molly, that disgusting comment, we handle it, then we talk about it," she says. "We feel through it. We talk about it with her."

Wait. She told Molly about the perverted comment? How did that conversation go?

"We sat her down like, 'Hey, Molly, listen. There are very broken people in the world, and they can be very, very, very disgusting-minded, and they're unsafe. They're not good people. And we've had several of these comments about you guys,' and she was like, 'Oh, my gosh.' And we're like, 'That's okay. We've got you. You are safe. We have you. We just want you to know that this is what's happening.'"

Jenn's parenting philosophy, she tells me, is openness and honesty. She and her husband endeavor to explain everything to their kids at a level they can understand. While in theory that has some merit, it's hard to understand what the benefit was of telling eight-year-old Molly that a stranger commented on a video wanting to see her underwear. What did she gain from that? What was the purpose of the conversation? Who did it serve? The conversation didn't serve as a jumping-off point to explain why Molly wouldn't be shown online anymore at all—she continues to be featured in content—so it's difficult to see how this level of transparency is to Molly's benefit.

I ask Jenn if she worries that her kids may, as they grow older, begin to resent her for involving them in family vlogging content.

Right now, she says, they love to be able to go back and watch their favorite videos and relive family memories, but Jenn is open to the possibility that they may feel differently one day. "And so if our kids, when they get older, come to us and be like, 'I friggin' hate that you did this to us in our childhood,' I'm gonna say, 'You are valid. And I am sorry that I did that to your childhood,'" she says. "I can be sorry, and the duality of life at the same time is I can still be okay with my decisions [about] their childhood. I stand by every decision, everything I've ever posted. Because for whatever reason, that needed to come out at that moment."

Knowing that you've shared details of your family's life in exchange for likes, follows, and brand deals is a "devastating thing that you have to live with," said Kathryn Jezer-Morton, a journalist and sociologist. "And you're constantly making up reasons why it's not devastating. That's what people are doing." That, it seems to me, is what Jenn is doing. Among online advocates for child influencer rights, there's the narrative that family vloggers and mom influencers are evil. They are viewed as monsters who have traded their family's privacy for money without a second thought. While I understand the need to separate the world into binaries and to categorize people as either fundamentally good or fundamentally evil, I just don't agree with this line of thinking. And neither does Jezer-Morton. "I think we do need to talk about it as these are human beings that have made human mistakes that are on a scale much larger than most of us would ever make, but they shouldn't be completely unrecognizable to us," she said.

It's not useful (or, I think, truthful) to see influencer parents and family vloggers as just unequivocally terrible people. "Yeah, I don't think they're monsters," Jezer-Morton said. "I think they're people that have made a horrible mistake and they're unable to reckon with their mistakes, and it's a tragedy." What is the fundamental mistake, in Jezer-Morton's mind?

"They wanted too much money. They did a thing that Americans do, which is that they decided that they wanted to become upwardly mobile, and they could do that. The opportunity to get richer, we're constantly inundated with images of what our lives would be like if we had more money. You know what I mean? No one is like, 'I have enough money and I love my life.' I think it's a horrible thing to do, and it's totally immoral. But I understand how people make that mistake."

The Jurgy family (@thejurgys with 257,000 subscribers on YouTube, 50,000 followers on Instagram, and 27,600 followers on TikTok) is made up of mom Nellie, dad Bryce, and their daughters Indie and Avalyn. When I call Bryce to talk to him about his family vlogging journey, I'm a little surprised that Nellie is on the phone too. I ask how they got started on YouTube and Bryce tells me it all began with a lump in Nellie's breast. At the time, Nellie was working at a law firm and Bryce was working at a creative marketing agency, and though he had floated the idea of YouTube and content creation before, Nellie wasn't into it. But then she woke up one morning and found a lump in her breast. Tests, scans, and a biopsy followed and, thankfully, it wasn't cancer—but the lump changed their lives anyway. "It was just like, 'Wow, life can change just like that,'" Bryce says. "With so many unknowns, Nelly looked to me and was like, 'Okay, let's do a YouTube channel where we adventure once a week. And that way you have memories, because we don't know what's going to happen with me. And it will push us to live life to the fullest.'"

Nellie speaks up for the first time in our conversation, after Bryce describes their first YouTube video in which they talk about Nellie's ongoing cancer scare (which was filmed before the results were in) and their plan to start their channel. "It's cringe watching now," Nellie says. I pipe in: "Everything is after a while, though." "Yeah," Nellie says. "That's so true."

They traveled to all fifty states in their RV, and their channel and platform grew. Soon, it became apparent that their vlogging had outpaced their old lifestyle: "Man, our jobs are holding us back," Bryce recounts. With ad revenue from YouTube and money from brand deals, they were able to quit their jobs and dedicate themselves to social media full-time.

Three years into their social media journey, their first daughter Avalyn was born, with the birth documented in a vlog titled "Our Baby Girl Is Finally Here!! Emotional Birth Vlog" (view count: 2.1 million). Near the 4-minute mark of the vlog, Nellie spreads her legs into a position that every mother who has ever attempted vaginal birth recognizes. "Bear down," the nurse instructs her, and she begins to push. As she pushes, the video flashes through footage of the couple's proposal and wedding and the moment they found out Nellie was pregnant. It really is a touching video, as it zooms in on Nellie's face before zooming out to footage of the couple dancing in front of a Mormon temple. Bryce is in the birth vlog and I find myself wondering who's filming it. "She has lots of dark hair," a nurse comments. There's a jump cut and then Avalyn, bloodied and covered in mucus and wailing, is placed on Nellie's chest. Nellie's legs are still open and she is crying too. The camera shifts to catch Bryce cutting the cord and separating mother and baby, and Avalyn wails those wild newborn wails. The top pinned comment on the vlog is from Bryce: "Hey new peoples! :) 1 - we love you and 2 - we just had our second baby! Birth video here: link."

During our interview, Bryce says that they originally created Instagram accounts for each of their daughters when they were born "because that's what people were doing. We instantly stopped. We announced the birth and then that was it. I'm like, they're not going to have their own Instagram accounts. That's silly. So there's definitely a fine line." When I check, Avalyn Jurgy still has her own Instagram page—it's private and has 2,959 followers. Her bio includes the link to her birth vlog and a link to the main family Instagram account. Avalyn's little sister, Indie, also has an Instagram—it's also private and has 837 followers. Her bio links out to all the other family members' Instagram accounts as well as their YouTube channel.

The Jurgys, like every family vlogger I spoke to in the researching and writing of this book, insist that they never force their kids to be part of their content. "If they want to be in a video with us, then that's great," Bryce says. "If not, it's no big deal." And though their social media career has allowed the Jurgys to quit their full-time jobs and open up an RV park, they maintain their motivations are pure; it's not about money. "Anytime we do post family content that our girls are involved in, the motive behind it is so that when someone sees it, they're like, 'Okay, I want to go spend time with my kids more.'" He continues, with reasoning that veers into the convoluted, to tell me that "even with the economy, good people with good intentions to provide a good life" sometimes are "not spending time with kids." He says some people have had broken childhoods and grown up in broken families, so he wants his family's videos to show people what a healthy family looks like.

"Maybe some child who didn't have that is like—" Bryce says, and Nellie jumps in, "that's what a family should look like."

"We're not perfect, but I think it's important. Personally, I feel like, right now—I totally get, like, we have to protect our kids, but we also have to share the beauty of what families are, because I feel like that's being torn apart in our world today," Nellie says, and I try to refrain from guessing her political affiliation. "The dynamics of families are being torn apart. We are trying to push that family can be awesome and that they are really important. There's a fine line of protecting your kids. But I also don't think it should stop you from sharing with others the beauty that comes along with having a family."

Okay. I'm not sure about Nellie's insistence that the idea of family is being torn apart, but I get what she's saying—that she wants to be a positive example in the world. And, Bryce adds, their oldest daughter, Avalyn, loves making content. "[She] loves to be a part of things. She loves watching it and laughing," he says.

This insistence—that the kids love being part of the content—is a common justification among parent creators. So common that in her master's thesis, researcher Bridie Hamilton outlined defenses that family vloggers and influencers often cite for their children's participation in content, with one of them being "But my kid likes making content!" Hamilton writes:

> Power dynamics between a child and a parent are inherently unequal, and children may feel pressure to comply with their parent's wishes or to seek their approval. Even if a child consents to be filmed, that consent may not be considered legitimate because children cannot fully understand, anticipate, or appreciate potential consequences. This is the case in social media content, where children cannot appreciate the scope or ramifications of their participation.

Later she writes:

> Even if a child genuinely enjoys participating in content creation, it does not necessarily mean that it is in their best interests. The law often protects children from doing things not in their best interest. Justifying the inclusion of minor children because they consented to their inclusion obfuscates the root of the problem.

Hamilton's reasoning makes sense to me—kids like a lot of things that aren't good for them. If I asked my nephew what he wanted to do all day, he would spend hours playing video games until he grew bleary-eyed. The fact that he enjoys video games doesn't justify his unfettered access to them. But what are influencer parents to do? They've made content creation their livelihoods, and some of their kids do want to be involved—though Hamilton's thesis makes me wonder if the kids *genuinely* want to participate in the family business. The Jurgys tell me their daughters aren't troubled by the more adult parts of influencing, like metrics and subscriber counts. Bryce says he never wants them to worry about how many views or likes something is getting. "Our kids don't even understand that," Nellie says. "They don't get that at all." Though I believe the Jurgys hold this to be true, I wonder if it actually is. How could a child who was literally born on YouTube not have an innate understanding of likes, followers, and subscribers? I worry they understand more than we can imagine.

Plus, the Jurgys tell me, "I like to think it'd be cool if I could see videos of my parents when they were our age as newlyweds. My most personal videos are [from] when [our daughters] were born, but watching those back are so magical," Bryce says. "It'd be awesome to see my parents' reaction when I was born and them talking about it and like, oh, look, there's me as a baby, a fleeting moment where I'm in the shot.

I think of that, like, *Okay, what would I love to see now of me growing up or with my parents and family?* But there's still a fine line."

(Okay. I get that. It really would be cool to be able to watch footage of the day you were born or the moments you took your first steps. But if a scrapbook is what vlogger parents are after, why does it have to be shared to millions? Families have made home videos for decades without sharing the content publicly. The logic falls apart like wet tissue paper.)

It's almost strange how closely the Jurgys' defenses for family vlogging adhere to those that Hamilton outlined in her paper. Hamilton even addresses the "digital scrapbooking" defense:

> Uploading for extended periods allows parents to store the modern-day equivalent of scrapbooks for their family lives in one place (Luscombe, 2017). If a digital scrapbook was the goal, parents could create private accounts or unlisted videos only available to close friends and families.

Anyway, the Jurgys tell me, they know that some family vloggers take it too far, like people who film their kids bleeding when their teeth fall out. But that's not them, they insist.

"I find it funny that there's a lot of backlash with family content . . . good family content is so needed right now," Nellie says, and her voice cracks as though she is about to cry. "I'm getting emotional as I'm telling you this because I'm seeing families fall apart, and I'm seeing the backlash on families. Honestly, just as somebody who does believe in God, I know that there is a power that's going against families right now, and it makes me more motivated. I will be so honest with you right now, this has been—doing YouTube is not easy. Doing Instagram is not easy. Putting your life out there is not easy. People are mean. I'm a very sensitive person, and when I read comments, I break down. But

I do it because I know that there are families and there are people out there who need family content. There is just that force of nature that is trying to make family not as important. I will fight that. I will fight that so hard because I'm sacrificing so much of my life to try to put a good message out there."

She seems to be crying in earnest now. "It's hard work," she says, "but we're doing it for the right reason."

Talking to the parents behind vlogging channels was fascinating. In every conversation, I found myself totally absorbed by what they were telling me. And here is what I have come to believe: To handle the psychological weight of monetizing their children's lives, these parents have to justify their actions. They're not just making monetized content; they're sharing their light with the world, they're raising awareness, they're showing the beauty of family life. Their kids, every single parent insisted, *like* being in the content. They're living the dream. I believe that they believe that. But I also think, whether it's in the middle of the night when they wake up to get a sip of water or early in the morning before anyone else is awake, that there are moments when they wonder if they're doing the right thing.

7

Bless This Brand:

Mormon Influencers, Trad Wives, and the Invisible Labor of Motherhood

In a commencement speech given at Brigham Young University–Hawaii in 2007, Church Elder M. Russell Ballard spoke of how the world had changed since he had been a missionary in 1948 and people received their news through newspapers and radio. "How different the world is today," he said, one of "cyberspace, cell phones that capture video, video and music downloads, social networks, text messaging and blogs, handhelds and podcasts." Addressing the soon-to-be graduates, Elder Ballard made a request. "Now, may I ask that you join the conversation by participating on the internet to share the gospel and to explain in simple and clear terms the message of the Restoration?" Among his suggestions: downloading videos from appropriate sites to share with friends, emailing reporters to request that they cover the church accurately, and starting a blog to "begin sharing what you know to be true." He concluded, "May the Lord bless each of you that you will have a powerful influence on those you come in contact with. The power of words is incredible. Let your voice be heard in this great cause of the gospel of Jesus Christ."

The members of the Mormon Church seem to have taken Elder Ballard's words to heart. Though only 2 percent of people in the United States identify as members of the Church of Jesus Christ of Latter-day Saints (colloquially referred to as LDS members or Mormons), the religion is overrepresented online; many of the top mom influencers and family vloggers in the industry are Mormon. Utah, the state that is home to the highest number of Mormons, is often pointed to by experts as the hotbed of family vloggers and mom influencers. But what is it about Mormonism that makes its devotees such successful influencers?

To answer this question, it's illuminating to return to the origins of the Mormon religion, which was founded by Joseph Smith in 1830 in Fayette, New York. "The religion itself is embedded in industrial capitalism," says Sarah Crabtree, PhD, an associate professor of history at San Francisco State University. "Mormonism and capitalism are like siblings, they grew up together." Because of that intertwining, Crabtree says, the Mormon religion highly prizes intense work ethic and fierce ambition—both of which are necessary to succeed as an influencer. As Joseph Smith shared the new religion of Mormonism through the mid-nineteenth century, the church looked for a place to settle and begin building a state they referred to as Zion, that would house the faithful and righteous people of the Mormon faith. Smith's missionaries spread across the country—and the world—to proselytize, and as they did, gossip spread with them.

It was whispered that Mormons were polygamous and one man could have multiple wives, leading to a growing distrust of the Mormon people. By 1838, the governor of Missouri declared that the Mormons "must be treated as enemies" and chased from the state. They moved eastward, putting down roots in Nauvoo, Illinois, which would become the Church's new headquarters. As the Church settled into life in Illinois, missionaries returned from their travels, bringing over converts from Europe. For a time, it seemed that Illinois might be where the dream of Zion could be realized. But the reprieve didn't last as reports of polygamy spread further and anti-Mormon sentiment grew. In 1844, Prophet Smith was arrested and then killed by a mob in Carthage, Illinois. Brigham Young became the leader and Prophet of the Latter-day Saints and led the Church west to settle in Utah. The early beginnings of the Church were rooted in a sense of persecution and otherness. Mormons, many Americans believed, were not true Americans—or true Christians. They were something different—something separate. Polygamy, which God had revealed to Joseph Smith as the path to heaven and eternal salvation, was judged as strange and un-American.

In 1904, after immense political and social pressure, the Mormon Church denounced the practice of polygamy (a move that many scholars credit for the Church's survival into the twenty-first century) and abolished plural marriage. "Once they finally ditch the polygamy, they have to prove that their monogamous heterosexual families are as perfect as anybody else's because they've been maligned for so long. There was this intense focus on the perfect household to help them assimilate and integrate into US society," says Crabtree. Their early history of persecution is the reason the Church became so invested in the appearance of perfection, according to John Dehlin, a member who runs the popular podcast *Mormon Stories*, which vows to help listeners grapple with hard truths about the history of Mormonism. "We were tired of being the laughingstock of America," says Dehlin. "We were literally targeted by the federal government. I think what we did [in response] is we became the all-American church." Perfection became an unofficial doctrine of the Mormon faith. And the focus seems to have worked. When we think of Mormons now, we often think of the perfect family: a mother and father and many beautiful children who are all dressed in color-coordinated outfits. In the Mormon religion, there is a premium placed upon the pursuit—and attainment—of perfection, especially in the family sphere.

That perfection includes journaling and record-keeping. According to Mormon doctrine, God told Joseph Smith, "There shall be a record kept among you . . . for the good of the church, and for the rising generations." Because of this dictate, the Church considers record-keeping a sacred duty. "The Lord has instructed His people to preserve and cherish their records and histories," writes Elder Steven E. Snow on the Church of Jesus Christ of Latter-day Saints website. "We gain strength from the stories of those who have gone before. We learn how to be strong and what brings happiness and joy in our lives." Young Church members are taught to keep detailed journals and maintain personal histories through means like scrapbooking (and eventually, blogging). "That's something that women are taught to do from when they're really young," says Kathryn Jezer-Morton, a columnist at *The Cut*. "They were early adopters of blogging because it was a way of telling their stories. Before the internet, they were really avid scrapbookers." The practice of scrapbooking was actually invented by Marielen Christensen, a Mormon woman, and to this day, many of the most successful scrapbooking companies are based in Utah.

Another pillar of Mormonism is proselytizing, or the spreading of the Gospel. Mormons are taught from a young age that every single one of them is a missionary. Benjamin Park, PhD, a historian at Sam Houston State University, says the Church was early to the realization of how the internet could be used as a new mission field. "They recognized how the new media landscape was evolving, especially with social media," he says. "These digital tools are now at your disposal, and they're coming through divine, miraculous intervention from God. You can use this digital medium as a platform to show how blessed you are and the wholesome values that you hold." Sam, the parent of a prominent vlogging family who used to bribe their kids with stacks of hundred-dollar bills to get them to take part in content creation, says that during their family's time as famous vloggers, at least fifty families joined the Church and told Sam's family they were the reason why. "There is a tremendous push from the Church itself and from its general authority and leadership to be public with your faith and your lifestyle and share it with the world. I think that's a big reason [so many influencers are Mormon]. We genuinely believe that we have something to share, like a light. You know, that's the most common word you'll hear Mormon bloggers and influencers say: 'We have a light to share with the world.'" What better way to do that than show your beautiful life with your beautiful home and beautiful children?

Speaking of beauty, Mormon women prize outward beauty and looking put together—or, as E. J. Dickson, a culture writer at *The Cut*, interprets it, "there's really no other way to say this—Mormon women tend to be really, really hot . . . It's because the Church puts a lot of emphasis on aesthetics and keeping yourself fit and wearing makeup at all times and being a good representative of the faith and of your family," Dickson says. "These are people who are very, very good at enhancing and marketing their own image." Dickson also points out that there are actual makeup tips on the Church of Jesus Christ of Latter-day Saints (LDS)'s website. "You are not required to wear makeup; however, wearing makeup can help you look your best," the website reads. In the Mormon Church, being beautiful is considered godly. A beautiful face, body, family, and house are reflections of the beauty of the love of God living within you. And to achieve that beauty, nothing is off the table: Researchers have found that 14 percent of Latter-day Saints had undergone some type of plastic surgery, compared with 4 percent of the general population. Plus, Dickson reminds me, Mormon theology dictates that members don't smoke or drink alcohol or caffeine. "That's another reason why they're so hot," she said. "They don't do stuff that inherently makes you un-hot."

So imagine: You have people whose faith is grounded in the pursuit of beauty and the spreading of the gospel, whose leaders are pushing them to make use of the tools of the twenty-first century to spread their faith even further. And they're raised to keep thorough records and journals of their daily lives. Do you see where this is going?

Mormons believe in the prosperity gospel, which is the concept that if you are faithful, God will bless you with material wealth. "Doctrinally, we believe that people who are righteous will be prospered," says Sam, the parent from a prominent vlogging family that we met earlier. "The Book of Roman teaches that God rewards you with prosperity, with money, with wealth, if you're righteous," says Dehlin. In other words: Making money is godly, and the more money you make, the more righteous you've proven yourself to be. And the influencing industry is incredibly lucrative.

It is common for Mormon people to get married young, due to a strong prohibition against premarital sex and emphasis on marriage and the importance of family and producing children. "The first commandment [God] gave to Adam and Eve was to "be fruitful, and multiply, and replenish the earth," reads the Church of Jesus Christ of Latter-day Saints website. "Those who are physically able have the blessing, joy, and obligation to bear children and to raise a family. This blessing should not be postponed for selfish reasons." And we all know there's nothing more lucrative for a mom influencer or family vlogger than a steady stream of pregnancies, gender reveals, birth vlogs, name reveals, and milestone photos. The algorithm loves babies.

The algorithm also loves perfect mothers. Mormon women are expected not only to be accomplished homemakers who have hordes of children but also to somehow still look more beautiful and put-together than the rest of us ever have. Mormonism is a heavily patriarchal religion, with a focus on men as the heads of the family who go out and work and take care of everyone financially while women stay back, keep a beautiful home, and raise babies. The focus on motherhood and domestic labor is part of what makes Mormons such successful influencers—and online influencing can be an avenue for these women to exercise their ambition in a way that is deemed appropriate by the Church. "Influencing is a way for women to work without challenging the patriarchal structure," says Dr. Park. "They can make money and participate in this ecosystem without leaving the home, without being seen as a threat to the traditional patriarchal structure of the faith."

Anyone who watched Hulu's hit reality show *The Secret Lives of Mormon Wives* saw its stars, young Mormon moms who rose to fame on TikTok, stating that they're all the breadwinners of their families. Fascinatingly, these women's success as mom influencers is predicated on selling a version of their lives that they're not actually living—one in which they are "only" moms, with no ambition, aspirations, or income outside of the home. But by being influencers, these women can both be what they were raised to be—doting mothers who keep beautiful houses—*and* exercise their ambition. "These women are monetizing their own lack of agency," Dickson says. "They get into these situations where they marry very young and have culturally been encouraged to have a lot of children very young. That is the lifestyle that is antithetical to their ambition. So they assess this and they look at this and they're like, 'Okay, so what's a way that I can fulfill my dreams and make money and fulfill my ambition while still maintaining this lifestyle that I've trapped myself into.'"

Natalie Jean Lovin, of the eponymous blog *Hey Natalie Jean*, was one of the original mom bloggers and grew up Mormon. She tells me the dictates of Mormonism perfectly align with influencerdom. "Mormon women are born and raised to be trophy wives, to be pretty, to be aspirational," she says. "If you're making money, God loves you. If you're beautiful, obviously God loves you. And so there's a lot of pressure for us to be beautiful and to put a lot of attention into how we look, how we sound, how we show up. And so it's tailor made for any media situation. We've all been raised from day one to be delightful and to be beautiful, and to be smart, and to know all these things, and to go to college, but not to actually need to use your degree because you're just going to have babies. But then when you have the babies, they're going to be the best babies ever. And you better as hell make it look amazing because your husband's going to come home, he's going to look at you and be like, 'What have you done all day?' That's the dark side of it, I think."

The Mormon Church plays a heavy hand in the success of their members as influencers. John Dehlin of the podcast *Mormon Stories* tells me the Church has a long history of providing resources to Mormon influencers. "The Church provides monetary support to influencers, brand deals, and book deals," says Dehlin. "They bring influencers to their headquarters in Salt Lake City and provide them with support and exposure opportunities."

Wait. What? The Mormon Church—which, granted, is the richest church in the US with an estimated net worth around $239 *billion*—provides monetary support to its influencers? What does that mean? I quickly text Shannon Bird, who was one of the OG mommy bloggers and raked in enough money for a few years that her husband quit his job (though he's back to work now that the heyday of mommy blogs is over and Shannon's success hasn't directly translated to Instagram), and she agrees to get on the phone with me. I tell her what I just heard, and she confirms that it's true. The Mormon Church pays its influencers. "Like directly?" I ask. "Actual money?"

"Yes," Shannon says. The Church reached out to her when her blog first started getting popular around 2011 and asked if she wanted to do sponsored campaigns for them. What would those campaigns look like? "It would be like, me doing charity," Shannon says. "Like, today we're giving to the homeless. So I would go out and get, like, rotisserie chickens." The Mormon Church paid her to give rotisserie chickens to homeless people? "Yeah," she says. "I think it was like an advertisement. Influencers are bigger than a billboard. I had a million people on my website a week. You know what I mean? Like, that's bigger than two missionaries going out."

The Mormon Church *pays* its influencers. I'm astonished by this piece of information. Shannon tells me that the posts didn't directly tie back to the Church—it wasn't like she was tagging them in blog posts or anything—but it seems like the idea was to pay her and other Mormon influencers to do things that would reflect well on the Church. It was an indirect commercial for Mormonism—the beautiful mom blogger giving out rotisserie chickens. In one of her Church-sponsored posts, Shannon holds several bags of groceries as her blond children swarm around her. "Sharing our experience the #lighttheworld calendar for those who'd like to help shine some light by serving others to those in need this Christmas season," reads the caption.

The Church asked her how much she would charge for a sponsored post, Shannon explains, but she felt bad charging her *church* so she gave them a lower rate. Back at the height of her fame, Shannon would usually charge at least $4,500 for a sponsored post. She asked the Church for $1,000. Once, someone in Church administration accidentally sent Shannon an email detailing how much they were paying other influencers. Shannon realized other Mormon influencers weren't charging the Church a special rate. They were charging their market rate; some people, Shannon remembers, were getting paid up to $8,000 for a sponsored post.

For the most successful Mormon influencers, the invitation to Salt Lake City to meet with Church Elders is a source of pride, Dehlin says. "If you're an influencer, it's a real honor to be chosen to attend those meetings. And the Church will often ask how they can support the influencers and provide them with guidelines for maybe what should and shouldn't be said." Shannon remembers going to those meetings. "They would provide a catered dinner at the Joseph Smith Memorial Building," she says. "It was just like mixing and mingling. It was business." At the influencer meetings, the Church would inform the gathered influencers about the upcoming campaigns they were hoping they'd take part in. "We'd all be there and learn our assignment," she says.

But working with the Church, Shannon warns, wasn't without its strings. Part of the deal was that she was encouraged to be upfront about her Mormon faith—and the advertisement of her faith seemed to invite criticism. "If I wore a bikini, it was like, 'I thought you were Mormon.' They come at you with pitchforks if they see your skin," she says. "Like, I'm sorry, but I'm sponsored by gyms. You're gonna see my skin, right? I don't believe in that. I don't believe I'm going to go to hell for showing my skin." Did the Church itself have a say in the kind of content she posted? Not directly, she says. "But you would get rewarded for being modest by getting campaigns." The Church did not respond to my request for comment.

For Mormon influencers, their faith and jobs as influencers become so intertwined that they are practically inextricable from each other, according to Sam, the head of a formerly prominent YouTube vlogging family. "There becomes this confusion between social media and their faith and their testimony," Sam says. If it's godly to be prosperous, and influencing is what allows for a family to be prosperous, then is influencing godly? The dedication to social media can take on an almost religious fervor. "They identify themselves with social media," Sam says. "And so for them to come out and speak against it would be like for somebody of religious faith to come out and deny God and turn atheist."

At the top of the list of Mormon mom influencers stands Hannah Neeleman, also known as @ballerinafarm, who has 10.1 million followers on Instagram, 9.8 million on TikTok, and 2 million subscribers on YouTube. She is beautiful, with long blond hair that is always either tied back haphazardly or perfectly framing her face. She is thinner than you might imagine a woman who has had eight children could be. She wears long prairie-style dresses that fall elegantly over her narrow frame. If she wears makeup, it's only enough to complete that kind of no-makeup makeup look for people beautiful enough to get away with wearing just a touch of blush and a swipe of mascara. In her content, her hands are never idle—she is either holding a baby, rolling out dough for a homemade baked good, or milking a cow. Sometimes, she is doing several of these things at once. (She's also a pageant queen. In 2024, Hannah competed in—and won—the Mrs. World beauty pageant just two weeks after giving birth to her eighth child, Flora Jo. In a *New York Times* story, Hannah says, "I am still bleeding a little" while getting her makeup done for the pageant.)

Hannah is fascinating. Thousands of think pieces have dissected her social media posts and the way she lives her life. Hannah, who was raised in a Mormon family as the eighth of nine children, went to New York City to study ballet at Juilliard when, a few years into her studies, she met Daniel Neeleman. Daniel, also raised Mormon, is one of the nine children of David Neeleman, the billionaire founder of JetBlue. In an article that immediately went viral upon publication in 2024, Daniel told *The Times* journalist Megan Agnew that when he saw Hannah at a college basketball game, he "was ready to go. Sign me up. I was thinking, *Let's get married.* But she wouldn't go on a date with me for six months." Hannah mentioned to Daniel that she was flying from Salt Lake City back to New York, without realizing that Daniel's father owned the airline she was flying on. Daniel made a call and arranged to sit next to Hannah on her flight. "And so begins their first date," the article in *The Times* reads.

> "Back then I thought we should date for a year [before marriage]," Hannah says. "So I could finish school and whatever. And Daniel was like, 'It's not going to work, we've got to get married now.'" After a month, they were engaged. Two months after that, they were married, moving into an apartment Daniel rented on the Upper West Side. And three months after that, she was pregnant, the first Juilliard undergraduate to be expecting "in modern history."

Henry, the Neelemans' first child, was born a week before Hannah graduated from Juilliard. Twelve years later, Hannah has borne eight children, and the couple lives on a 378-acre farm in Utah, where there is no ballet to be found (the space that was meant to be a ballet studio was later turned into the kids' schoolroom). Not only is Hannah the ultimate Mormon mom influencer, but she's also the Trad Wife to End All Trad Wives. (Her next closest rival, Nara Smith, who famously makes her kids snacks like graham crackers from scratch, is tight-lipped about her religious leanings but was married in a Mormon temple.) Let's back up for a second and define *trad wife*, which refers to a "traditional wife" or a woman who practices conventional gender roles in her relationship and basically does everything that we expect the "ultimate" wife would do: cook, clean, bear children, all with a smile on her face. Trad wives are, it seems, suddenly everywhere. Scroll for a few seconds on any app and you'll see a beautiful mother baking homemade bread with her children walking in and out of the frame behind her. Trad wives sell a fantasy world in which one income is enough to sustain a family and all a woman has to be responsible for is taking care of her children and maintaining a clean, aesthetically appealing home. They also present a version of the past that they presume to be more natural, one in which gender norms are reinforced as a husband goes out and works and a woman stays home.

Part of the controversy of the trad wife movement lies in the lack of choice and agency: Did women who were "trad wives" in the past really choose that? Or was it the only option they had? It was only post–World War II that this version of the family—in which a man worked and a woman stayed home—was created, says Sarah Crabtree, the history professor who helped orient us earlier in this chapter. "The bargain was that, after World War II, you paid working-class men enough that they could keep their wives home because women had worked during the Depression," she says. "The grand bargain of neoliberalism was the family wage. Then the economy began to collapse and they took that away. And the way the economy was propped up was by allowing or encouraging women to go to work, so suddenly families had two incomes, not one." Though people were making less money, Crabtree explains, the economy was bolstered by cheap goods from other countries. But in the 2020s, not only are people making less money, but the family wage has disappeared, and goods aren't as cheap as they once were. "It's like they've been moving the cups around, and it turns out now that you lift up all three cups and there's nothing underneath any of them. That's this very precarious, scary moment we're in. And so I'm not surprised people are trying to reconstruct the past."

The idea that trad wife content is surging because of economic and political precarity makes sense to me. The present is so scary—why not return to an idealized past? In fact, that's what one study found. "Tradwives leveraged maternal identities during pregnancy as a push factor toward their communities with a central focus on motherhood and, equally, a level of trauma or instability in childhood, and/or a drastic change of life circumstance [i.e., death, health crisis, pandemic] was construed as a touchpoint to encourage a turn to traditional roles." Millennials are the first generation who are expected to be less financially well-off than their parents; many of us consider traditional markers of adulthood, like home ownership, a pipe dream. Though the cost of living continues to balloon, wages remain stagnant. I understand why, amidst this uncertainty, the allure of trad wifedom gains traction. For young women, it's a way of stepping out of the hamster wheel of Girl Boss *Lean In* Feminism, the pressures of the job market, and the anxiety of trying to make your way in a world in which a white woman makes 83 cents and a Black woman makes 69 cents to every $1 a white man makes.

I understand the appeal of the trad wife role. It seems like a relief to decide you're done fighting in this society, where the wage gap persists, reproductive rights are vanishing faster than we can fight the forces that are expelling them, and post–#MeToo backlash seems to have squashed any gains the movement ever made. It feels impossible to both nurture a career as a young woman and get married and have a family, and if there's an option to step back from the crushing pressures of at least one of those struggles, I don't know that I can blame anyone for taking it. Trad wife content promises an alternative: a soft life in which your only worries are taking care of your sweet babies, baking loaves of bread and cooking meals to nourish the family, and tidying your house by the time your husband gets home from work. You don't have to struggle, Mama. You don't have to let someone else raise your babies. You don't have to miss their first steps because you are sitting at a computer, plugging away at a job you don't really care about. You don't have to worry about your career trajectory or quarterly reports or paid time off. You can rest, Mama. You can make the home your domain. And you can make it beautiful—heavenly, even.

I understood the draw of trad wife content more before I became a mother myself. I would scroll through TikTok and Instagram and spot these beautiful women with their perfect balayage hair soothing their babies and making three homemade meals a day, and it looked so peaceful. Their job was to take care of their children and their homes and in return, *they* were taken care of. It seemed like a promise I could imagine someone would willingly make.

Then, I became a mother myself and I saw the truth behind the polished social media veneer. Motherhood can be rewarding and joyful, but it is also full of irritation and drudgery, dirty diapers, sleepless nights, and moments where you wonder despairingly if this is what your life will look like forever. Motherhood is blood and guts and the physicality of a body pushed to its breaking point without the relief of actually breaking. This seems obvious, but it's worth pointing out: The reality of motherhood is nothing like it appears on Instagram or TikTok or YouTube. The labor of motherhood is endless: There are babies to feed, milk to pump, formula to mix, laundry to fold, dishes to clean, meals to cook, books to read, tummy time to practice, naps to adjust, and bedtime routines to create and adhere to. There are clothes to sort, diapers to throw out, baths to give. Messes that must be cleaned, over and over.

The labor of motherhood is also, mostly, unseen. It occurs privately, within the home. It is not glamorous and it is not paid. It is expected, taken for granted. Invisible. The promise of the mom influencer and family vlogger is a dream: You will be paid for the labor of motherhood, and you will be paid well.

Right? Well—kind of. That's the thing about mom influencers—they're not getting paid for the labor of motherhood because there is no payment for the labor of motherhood. They're getting paid for the *performance* of that labor. "They're getting paid to look a certain way, for their house to look a certain way, for their food to look a certain way, for their kids to look a certain way," says Sara Louise Petersen, author of *Momfluenced*. "They're more making money off of their styling abilities, their marketing abilities, their writing and video abilities, and their cohesive branding. It's not that they're making money off doing the dishes. It's showing how doing the dishes can be a life-changing moment or whatever."

This is the fundamental disconnect that I, and maybe you, may see as disingenuous or frustrating when it comes to mom influencers and family vloggers. When you see their content, it's easy to think they've figured out a way to make the drudgery and chores of daily life magical. But they haven't—they've simply monetized the performance of that drudgery. Nara Smith isn't getting paid to make bubblegum from scratch. She's getting paid to set up her phone and put on a perfect dress and curl her hair and perform making bubblegum from scratch. Hannah from Ballerina Farm isn't getting paid for taking care of her eight kids. She's getting paid to perform motherhood in front of a camera and edit out the ugly parts. Because there is no money in the unpaid, and often unseen and unappreciated, labor of motherhood. It is, by definition, left out of the economic system.

There's a part of me that wonders whether the backlash against mom influencers and family vloggers is actually grounded in misogyny. Through a certain lens, the content mom influencers create could be seen as a feminist act to monetize the unpaid labor of motherhood. Jo Piazza, author and host of the podcast *Under the Influence*, sees it that way. "It absolutely could be a feminist act to show the labor of mothering, to pay mothers for their work—albeit in a roundabout way through the shilling of products—and as a way to show mothers that their work has purpose," Piazza says. "This is a multibillion-dollar industry run by women for women. It is a place where women have agency and power."

Petersen sees it as more complicated—there are feminist aspects to mom influencing, but it's not cut-and-dried. "Talking about the labor of motherhood publicly and amongst ourselves is really—and should be—a feminist goal," she says. "I just don't know that its relationship to capitalism can be firmly a win." And I get that. Is shilling #sponcon to other moms really a feminist act? Maybe not. But what about the act of creating profit out of work that has typically been so underappreciated and, historically, unpaid? Maybe.

It's not a coincidence that most family vlogging accounts are at least outwardly headed by the mothers and that #mominfluencers are so much more popular than dad influencers. "It says quite a lot about the continuing and unresolved difficulty of caring for children and earning a living at the same time as a mother compared to a father," says Lyn Craig, a professor of sociology at the University of Melbourne who co-wrote the paper "On Digital Reproductive Labor and the Motherhood Commodity." "Being an influencer on the internet is something you can do from home, flexible as to time, and so conceivably, is something you can do with your children in the house or even in the room. But it's a way of doing it while managing your care responsibilities, which for many women, but especially with young children, is the primary thing they have to do. And then they have to fit their work around it quite creatively. It's amazing, the adaptation. And the opposite is more common for men, where they fit their care around their work."

It's so simple and so true—men fit the care of their children around their work while women fit their work around the care of their children—that it makes me dizzy. This happens even in my marriage, with a husband who identifies as a feminist and didn't mind when I didn't take his last name. He is able to fit the care of our child around his work; I, mostly, fit my work around the care of our child. So perhaps it's obvious why women become mom influencers or start vlogging channels: It is a way to meld the work of making money with the work of being the primary caretaker of their children. We now live in a world where every part of our lives can be commodified. People make thousands of dollars live-streaming themselves cleaning their kitchens or organizing their pantries. Why should the labor of motherhood be exempt from the same commodification?

In the paper "'Mommy Meets Money,'" gender studies researcher Erla Gjinishi examines the biopolitical production of life as a marketable commodity. Gjinishi tells me she became interested in the topic after noticing the "blurring of the boundaries between productive and reproductive labor." She began to understand the mommy blogger through the lens of Marxist philosophy, where all of life exists under the logic of the capitalist market. "We are in this almost breaking point in the history of capitalism, where it becomes so absurd what you can buy and sell," she tells me, and I think about the TikTokers who make hundreds of thousands of dollars a year making ASMR videos or live-streaming themselves completing their chores. "Now, every single detail of life can be marketable and commodified as long as you do the right packaging to sell it. Women have been doing unpaid labor throughout history. But what we are seeing now, with the rise of the mommy blogging and mommy Instagrammers, is how much work is done and how costly it is, emotionally and mentally, for women."

But, Gjinishi takes pains to point out, there is a certain kind of woman who is best suited to monetize the labor of motherhood: mostly white women living in the West. And there seems to Gjinishi to be a level of trickery involved, where the blogger or creator is purporting to simply be sharing the experience of motherhood to help the viewer when really, the creator's intention is to commodify the performance of the experience and make money out of it (though this trickery is endemic to most influencers, not only mom influencers and family vloggers). As I'm writing this, I pick up my phone and open Instagram, where a mom influencer has posted a story about how difficult it is to stay hydrated as a mom. I click to the next story, where she links a commissionable hydration powder.

Gjinishi says it disturbs her, the idea that every single part of life can be packaged and commodified and sold. But there's also a part of her that sympathizes with the people packaging their lives because they live within a system that demands constant commodification. "This is the global economic order and we have to participate in it. There's no way out of it," she says and is quiet for a moment. "So I think there's a bit of hopelessness, maybe."

I know what Gjinishi means—can we really blame the people finding a way to work within a system they didn't create and have no chance of changing? Why do we direct our ire toward the mom influencers and family vloggers? Even in asking this question, the answer is already evident: because they're available to blame. We can't blame the people who created the status quo, but we can blame the people who benefit from the exploitation of it.

Let's contemplate what Amanda Montei, author of *Touched Out: Motherhood, Misogyny, Consent, and Control*, says: The work of the mom creator should not be viewed as inherently feminist or antifeminist but instead as work that upholds the status quo, "particularly when what you're doing is putting on a performance that is furthering often misogynist standards that then [other] women and new parents are holding themselves up to. They are perpetuating a self-sacrificial, saintly image of motherhood. And that's not new. That's always been an unrealistic standard of motherhood. So it's not disrupting anything. I wouldn't say that's feminist. I wouldn't say that's beneficial." Making money is not in and of itself a feminist act, especially when it's money made off the backs of convincing other women their motherhood and family lives aren't good enough and that, if they just got their kids a big enough Christmas haul or bought a specific hand soap (#sponsored), they might be the mom and the woman they hope to be.

While perhaps it's not a feminist act to create paid labor out of unpaid labor, it is an act of agency within the confines of the patriarchy. It's why so many mom influencers and family vloggers are Mormon. There are only so many acceptable ways to be an ambitious woman—influencing is one of them. There's a reason that there is a high correlation between people who find meaningful careers as mom influencers/family vloggers and people who started having kids when they were young themselves. Many of these women didn't have the chance to become anything else outside of the patriarchal mold: They transitioned from girlhood to wifedom to motherhood without the luxury of self-actualization or developing a standalone identity as an adult. As mothers, there were only a few ways for them to exercise their ambition while working within the enclosures of what is considered acceptable to their communities—and, frankly, to the world. This is the work and the ambition they're allowed within the confines of a capitalist patriarchy. And if you squint, it does look feminist; at the very least, it's a way of being an ambitious, capable, striving woman while still centering what people most expect you to be—a mom.

The trad wife performs an imitation of the past because the present is so uncertain. She drapes herself in prairie dresses and flowing aprons, smiles into the camera that captures her every move even as she insists she's not working at all. The mom influencer props a phone on a tripod stand and performs motherhood in front of it, links to the products that she swears make it all just a little bit easier, Mama. The viewer observes from her stained couch, dressed in sweatpants she's had on for three days. She can't help but watch and click and buy because the thing she wants most is what the moms inside her phone are really selling: the promise that there is meaning and beauty to be found in motherhood. All she has to do to get there is like, follow, and subscribe.

8

From Content to Consent:

The Parents Who Changed Their Minds

Bethanie was only eighteen years old when she got pregnant and married in 2013. Her husband was working nights at UPS and she was lonely during her pregnancy. One night, she logged on to WordPress and started a blog called *The Garcia Diaries*. The blog is suspended in time. The About section tells the reader that Bethanie is still married to her ex, Anthony, who she calls her high school sweetheart. In her first post, she introduces herself and her children: "My name is Bethanie Garcia. I have a nine-month-old daughter named Brooklynn Alessandra and am 6 months pregnant with a second daughter, Harlym. Close in age, I know." As I scroll through her blog, I see a post on Brooklynn's first birthday party, advice for what to do when you can't recognize yourself in the mirror anymore (yes, I clicked), and a recipe for bacon-wrapped chicken (I didn't click). There are posts detailing her son Deuce's path to neurosurgery and his healing, tips on how to get your toddler to nap, and six ways to style off-the-shoulder tops.

Today, thirty-year-old Bethanie, who is now separated from her husband and engaged to someone else, tells me the blog was "such a good creative outlet," but it was also a vehicle for income. Just a few years after starting *The Garcia Diaries*, her ex quit his full-time job to become a stay-at-home dad so Bethanie could focus on her burgeoning blog. Being her family's breadwinner wasn't something she expected as a young mom without education past high school. "The fact that with no college education and with five children now, I can support my family, it's truly wild and a dream come true, and I never could have possibly imagined it at all," she says. Though she doesn't post on her blog anymore, she has nearly 340,000 followers on Instagram, where she posts daily—and sometimes hourly—updates on what she's wearing, what she's eating, and what her kids are up to. Bethanie is astounded by her own success and grateful for the freedom her career as an influencer has given her, but she also worries that she has shared too much. If she could go back, she says, she wouldn't have ever shared her children's names or faces. But now, there's ten years' worth of content online, so what is she supposed to do? She can't change what she shared in the past (though she could choose to stop adding to it now).

"I've had situations where I'm like, 'Okay, this doesn't feel safe,'" Bethanie says. "For example, I was in the toy section at Target with my daughter in the cart, and my son was in the back part of the cart just sitting in there. I turned around for a second to look at something. My hands are still in the cart. While I was turned around, someone came up and was like, 'Deuce, is that you?' He was three at the time, or maybe four. It didn't really faze him because he was a toddler. But it fazed me a lot because I was just, like, oh, my God. I jumped to the front of the cart to stand in between him and the person. I think they meant well, and they were just excited to see someone that they follow closely online in person. But for me, it definitely gave me pause. So that doesn't feel safe to me."

Bethanie goes back and forth "every single day" on whether she should keep sharing her children. "I think I know what the right thing is deep down, but I'm also so proud of my family and want to show them off," she says with a self-conscious laugh. "I need to stop, probably."

Bethanie hasn't stopped sharing her kids online but in recent years, a growing number of influencers have. One of those parents is Maia Knight, a twenty-eight-year-old mom who rose to viral TikTok fame for sharing videos of herself single parenting her twin daughters. In early videos, Maia held an infant in each arm while preparing their bottles of formula. Each morning, she posted a video of herself running up the stairs to get the babies from their cribs, and viewers soon became attached to both Maia and her daughters Violet and Scout. (There was a long-running joke about the way auto captions would butcher Violet's and Scout's names that turned into fans calling them things like Scotch & Vodka and Silence & Violence and Salt & Vinegar. At one point, Maia even released merch with the nicknames on tote bags and phone cases.) At the height of her TikTok fame, Maia had 8.1 million followers who tuned in to see what daily life was like for a young single mom of twins. I was one of them. When Maia came up on my For You Page, I was immediately drawn in. It was interesting to see a young mom handling twins, for sure, but there was something else. Maia most often appeared on camera in a sweatshirt and sweatpants, with her hair tied up in a bun. She looked happy and full of love but also stressed and exhausted. It seemed real in a way that other mom influencer content did not. And Scout and Violet were *so cute*.

Then in December of 2023, Maia posted a video announcing she was no longer going to show her daughters' faces online. "I have been talking about and taking actions towards taking the babies off of social media for a year now," she says, addressing the camera with her hair in its signature messy bun. "They're toddlers now, and I have decided to not show them anymore. I'm not being forced to do anything. I'm making a choice for my daughters to protect them. I'm not, like, taking a big stance about showing your kids or not online, I'm just doing what's best for me. It's not that complicated, it's not that dramatic." She goes on to say that she knows she'll lose followers but hopes that the majority of "normal people" will understand her decision and support her regardless. And she had a message for detractors: "If you feel that strongly about a stranger that you've never met and the decisions that I'm making for my kids, maybe you should go to therapy." From then on, Maia never showed the girls' faces again. If they were in her videos, they were either filmed from the back, seen from the side, or emojis were put over their faces. This is what Dr. Michael Walrave, a professor of communication studies at the University of Antwerp, calls "mindful sharenting," a practice that he says has become more common. "There is a social pressure on parents to show their performance as a parent online," Dr. Walrave says. "Mindful sharenting—which can include photographing a child from the back or putting an emoji over their faces—can be a way to navigate the perceived risks of sharing your child online while still showcasing your experience of parenthood."

As of this writing, Maia has 7.7 million TikTok followers, only a slight decline from her height of 8.1 million. The reactions to Maia's decision were varied. There were people in her comments begging her to show her kids again. Fans called themselves Violet's and Scout's "TikTok aunts" and wondered how they would be able to handle not seeing them anymore. They said they were devastated, they were angry, and they missed the babies already. Other commenters questioned the timing of the decision. If Maia was so worried about the girls' safety, then why had she posted them in the first place? And wasn't it convenient that she was now deciding to take them offline after building a platform of millions by using them? Even if she wasn't showing their faces, some said, it was still exploitation to feature them in her content at all. Still others straddled a thin line: They applauded Maia's decision while at the same time denigrating her for making it at a time they considered "too late." Fine, these comments said, she was doing the right thing. But she was still a monster for ever having done anything else.

Conspiracy theories abound on TikTok about Maia's decision, with most theorists confidently proclaiming the reason Maia took her kids off of TikTok is because their dad took her to court and obtained a court order dictating that she's not allowed to show them online. Though there are thousands of comments and videos that claim this, there is no evidence to prove this is the case. Maia did not respond to my request for comment. The decision to take her daughters offline seems to have been hers. And her influencing career has survived it: One source familiar with the income of influencers estimates Maia makes between $3 and $5 million annually.

Aspyn Ovard—who has 2.4 million Instagram followers, 2 million YouTube subscribers, and 1.1 million TikTok followers—rose to fame in the early aughts as a high schooler posting vlog content and makeup hacks before transitioning into a bona fide mom influencer after giving birth to her first daughter in 2019. For four years, Aspyn's daughters were the center of her content, smiling into the camera as their mom vlogged their daily routine or grinning in a post #sponsored by Walmart. That changed in June of 2023 when Aspyn posted a photo with the caption "family beach weekend :hearthands:." In the photo, she and her then-husband Parker Ferris are each holding one of their daughters, and small brown emoji hearts are placed over the girls' faces. Since that post, Aspyn, who previously shared much of her pregnancy and motherhood journey, has not shown any photos of her daughters' faces (her third daughter was born in 2024). In a 2024 TikTok, Aspyn said her biggest regret when it comes to social media is ever showing her kids online. "If I could go back in time . . . I would not share my kids in the way that I did. I would not share their legal names, just for their privacy. I wouldn't have shared any updates specifically about them and their personalities. I wouldn't have shared so many pictures of their faces. That is my one regret that I'm like, dang, I really wish I could go back and do things differently." Aspyn explains that she's gone back to delete old content featuring her kids, but when it comes to sponsored content, she hasn't been able to, because featured in some of the contracts is the understanding that the content will live on her platform indefinitely. If such content is removed, it could be considered a breach of contract, according to Casey Handy-Smith, a lawyer who counsels creators. "The consequences depend on the agreement, but it could mean refunding part of the fee, paying a penalty, or other remedies spelled out in the deal terms," Handy-Smith says.

As in Maia's case, rumors abound regarding why Aspyn isn't showing her kids on social media anymore. The rumors range from wildly speculative to mildly possible, but ultimately there is not enough evidence to support any one over the others. On Reddit, anonymous users are convinced that the reason Aspyn took her kids offline is because one of them is terminally ill (though there is little evidence for this theory). But I think the fact that these rumors exist illustrates something important about what it's like to put your kids online: Even when you take them off, you can't truly take them off.

Over on the r/aspynovardsnark subreddit, which boasts 34,000 members (more on snark communities coming in a later chapter), anonymous users post screenshots of videos or social media posts of family members in which they swear they can see a glimpse of one of the children's faces. Fans (and haters, though they prefer to call themselves "snarkers") have formed parasocial relationships with the kids that don't end simply because they're not being shown anymore. It reminds me of something a mom (who requested to remain anonymous) with millions of followers who took her kids offline told me about her decision: "Everyone's reaction proved to me that I did the right thing."

When I search for *Barefoot Blonde*, the blog that propelled Amber Fillerup Clark into the influencer stratosphere, I get a 404 error. "The site you were looking for couldn't be found," the screen tells me. Clark deleted it, but not before it made her famous.

Amber started *Barefoot Blonde* in 2010 when she was twenty years old. On the blog, she shared her adventures (and misadventures) in travel, dating, and beauty school. The blog followed her as she met her husband and got engaged, then married. It chronicled her pregnancies and the infancies of her children and the family's move to New York City. Amber tells me she shared about "just everything, honestly," so when she started her motherhood journey, it only made sense to share that too. Plus, she didn't want to leave her kids behind when she had to go work creating content. "It felt very vain to go get in an outfit and leave my kid to go take outfit pictures," she says. "I had a really hard time wrapping my head around that. So that's when I started to be like, 'Well, what if I wear my outfit, but I bring my kids with me? We just go have fun together and document it that way.' So I think in the beginning, it started off very innocently, just me wanting to spend time with my kids and not having to leave them."

Telling me about her early success, Amber is humble, as if she just stumbled into being one of the foremost mom influencers of the 2010s. "I think it was just right time, right place," she says. Her blog started getting popular right as Instagram and Pinterest exploded, she explains, and her content went viral easily, leading to sponsorships and brand deals. "It was just one thing after the next, where I would start to get these opportunities. And I was so excited about it and having fun. So, I just followed it."

Back then, Amber's kids, Rosie and Atticus, were featured in so much of her content. Scrolling back through the framed squares of

Instagram is a window into that time in her family's life. There they were, relaxing as a family on a lazy Sunday. There they were, advertising Seventh Generation–branded diapers while picking berries. There they were, dressed in perfectly matching Halloween costumes. Amber tells me, "I have really great memories of that time period of life. I thought that was the coolest thing ever. I felt like I hit the lottery doing that. You know what I mean? It's hard not to just look at that as the coolest thing."

In 2017, *The Atlantic* profiled Amber in a story titled "The Enviable Life of Instamom Amber Fillerup Clark," the end result of which she tells me she found "devastating" at the time. When I read the profile, I can see why it upset her. Her motivations are portrayed as shrewd and calculating:

> Fillerup Clark rejects the idea that she whitewashes motherhood. "We take pictures as it happens. Whatever we get, we get," she said, as she winnowed about 30 photos of her kids in their Trump and Clinton costumes down to six blogworthy shots. She noted that she regularly shares aches and pains in the text accompanying her photos. And when it comes to her own appearance, she is candid about the ways she gives Mother Nature a helping hand, openly discussing her fondness for sunless tanning, false eyelashes, veneers, and hair extensions . . .
>
> "So we're thinking of having an indoor gym in our home because if we could even say yes to one or two fitness campaigns, then that would pay for the gym itself," Fillerup Clark explained. They'd sprung for an outdoor shower for similar reasons. "Sometimes we'll have a campaign where we're doing shaving cream, and it's a little awkward to be indoors in your shower, so it makes more sense to have a beautiful outdoor shower and do it out there." They were incorporating pic-

turesque window seats, and had come up with a special design for what they called "Amber's hallway": It would be extra wide and lined with windows and, according to Clark, was partly "based off of 'I want to take pictures there.'"

"The more our house becomes Pinnable, the more it leads back to the website," said Clark. "We want it to traffic well. We want it to go viral."

The publication of this story was the beginning of Clark's decision to take Rosie and Atticus offline. At the time, Amber tells me, she was defensive. "I was like, well, yeah, of course that's my job to want engagement. I have to think about that. It's literally my livelihood." But the more she kept thinking about it, she wondered: "Is that the type of person I am? Do I actually want to go about life looking for viral moments?"

Amber tells me she was careful to never share things about her kids that she deemed too personal. Off-limits topics included tantrums, poopy diapers, and meltdowns. And she was panned for that curation, accused of broadcasting a perfect image of motherhood in which children are always well behaved and coiffed. But if she had shown the tantrums or potty training, she would just as likely have been raked over the coals for that too. Which makes me wonder: Is there any right way to be a mother on the internet—or off it? Amber was aware of the accusations that she was platforming unattainable perfection, but that was—to be frank—her job.

When she decided to take the kids offline entirely, it started gradually and then happened all at once. After *The Atlantic* story, she started to notice the way people were talking about her kids in the comments of her social media posts. "People were just such assholes," she says, and it's the first time she curses in our conversation. "Nothing is off the table for people to be an asshole about, including kids. And I had the thought of like, *Oh my gosh, I'm putting my kids in a situation where people could be mean to them.*" When her son Atticus started school, his teacher told Amber she followed her blog. Amber wondered how it would affect real-life Atticus to be measured up against the picture-perfect version of him that was broadcast online. "They are still kids, and they might have big feelings at school, and I don't want them to feel like 'Wait, but you're so happy on your mom's account' [and] be held to a different standard than any other kid would be. I just wanted them to have a normal school experience, no preconceived notions about them. No expectations. Just grow up normal, I guess."

By the time she was pregnant with her third child—a daughter she would name Pepper—it was 2022, twelve years after she first started her blog. The feeling of unease finally crystallized into action. She deleted her blog and took her YouTube videos down. When her children were in her content, Amber took photos that were artfully blurred or featured only the sides of their faces or back of their heads. I ask: Was there a single moment or comment that pushed her over the edge? No, she says. "I think it was more just a gut feeling. Of knowing what was right for our family." She knew she would lose followers, and she says she did, but the amount ended up being negligible.

Amber was among the first—and possibly *the* first—big mom influencer to take their children offline. Though it's become more normalized for content creators to protect their children's privacy, her decision stood out back in 2022. It also let her return to her roots. "I think if anything, I've actually gotten back to what I started with, which was just sharing my personal life and my personal experience, my thoughts, my travels, whatever."

Though Amber is trying to protect her kids, they don't understand that yet. "They ask a lot, like 'Why can't I be in this video if you're filming an ad or something?' And then I'll have to explain it to them." This makes me think of the many influencers who have told me their kids love being in their content. If the kids want to be part of it, how can you say no? How does an influencer mom explain to Gen Alpha kids that they can't be featured in her content? "I honestly just say, 'One day you'll understand. One day you'll get it. But for now, you'll just have to trust that I'm making the best decision for you.'"

Amber tries to find different ways to let her kids stay involved. If a brand asks that she include her kids in her deliverables, she offers alternatives. For a recent campaign with Simple Mills, she showed herself making the kids an after-school snack without showing the kids themselves. In a magazine feature, her husband took photos with long-exposure where Amber was standing still and the kids were running in a blur around her. It still captured a certain depiction of motherhood but also preserved the kids' privacy. "There is more room for regretting decisions when sharing kids than if you don't. You'll never regret not sharing kids, but there is a lot of room for regret or resentment when kids are older, and I just didn't want to leave room for any of that."

Felix, who uses they/them pronouns, and Morgan, who uses she/her pronouns, are the couple behind @couplagoofs on TikTok, where they have 528,500 followers, and Instagram, where they have 125,000 followers. They're a queer couple who first went viral with a video in August of 2020 where Felix is filming Morgan crying. "Morgan, why are you crying?" Felix asks through laughs. Morgan is laughing too but crying at the same time (as so many of us were in 2020). "Because I want to go to Texas Roadhouse," Morgan says. In an interview with the couple, Morgan tells me the video was more about "how stressful everything was and how I wanted things to go back to normal" than the desire for Texas Roadhouse (though she did also want that). The couple gained 10,000 followers from that video and kept making content as they moved to Oregon and got married and started their trying-to-conceive (#TTC) journey. In a video viewed 1.5 million times titled "How We Got Pregnant at Home," Felix describes what they call the turkey-baster method to queer pregnancy. Morgan thinks their path to parenthood—through a known donor and a syringe at home—made them more accessible than queer content creators who can afford in-vitro fertilization, which can run up to $30,000 per cycle. "The content we're making is a little bit more accessible and attainable to people," she says.

The couple's #TTC and #queerpregnancy journeys were heavily documented on TikTok and Instagram. The announcement of their child's birth came in a reel where the baby's face is covered by a sun emoji, and every photo and video they've posted since had some kind of illustration over the face of their child. On July 19, 2024, about a month after the birth (though I can't be totally sure because Felix and Morgan never said exactly when their baby was born), they posted a photo of the three of them on Instagram with the caption "Ways We're Keeping Our Baby Safe on the Internet." The caption outlined their plans including: (1) only referring to their baby using fake nicknames starting with the letter *G* (from a scroll through their content, I see them calling him Gonzo, Gemini, and Gluten); (2) not showing his face at all by using a custom bear "privacy cover" emoji over it; (3) centering their experiences as parents rather than their son's experiences ("for example," the caption reads, "it wouldn't be 'the baby pooped,' it'll always be 'I changed a diaper'"); and (4) purposefully "misleading" followers about certain information like his birth date.

Does it really matter if the internet knows the date of a child's birth and their full name? Bridie Hamilton, an ethicist who received a master's from Duke University in bioethics, technology, ethics, and policy, thinks so. While studying for her master's degree, Hamilton became interested in the ethics of sharenting. At what point, she wondered, did posting about a child cause them actual harm? Hamilton and her advisor set out to determine the salient characteristics of what would be considered a privacy harm to the posted child. "That would be something like filming regularly occurring for a prolonged period of time on a very predictable schedule or something like work being very indistinguishable from play, or emotionally volatile or misleading titles of the content," she said. "[There are also] things that we would consider a bit more mundane, and it might not even matter in the long run, but things like full names, birthdays, heights, that if you were paying attention for long enough, you could reasonably impersonate that person and commit fraud on their behalf. You could figure out where they went to school or whether they had cheerleading practice. So that privacy intrusion could materialize in some legitimate threat or harm."

But when making decisions about their children, most parents aren't thinking about the academic research that illuminates potential harms. So what were Felix and Morgan thinking about? They felt very affected by a video they saw from another creator whose toddlers were recognized by a stranger in a grocery store. Though the couple was still on their #TTC journey then, they realized they wanted to protect their future child from something similar. "Even with just having a trying-to-conceive journey, people feel very comfortable asking incredibly invasive questions and also giving a lot of unsolicited advice," Morgan said. "I think that in general often comes from a place of care because they feel connected to us. Because we're characters to them."

We're characters to them. This stops me in my tracks in both its simplicity and its truth. Influencers are characters to us, like sitcom characters were in the past—people whose families we watched and derived entertainment from. But behind every influencer is an actual person, like Morgan or Felix, who is living an actual life.

What they want for their son is for him to be able to grow up without strangers forming a parasocial relationship with him, Felix says. And, Morgan adds, without people trying to touch him at the grocery store (which has happened to the couple themselves). Plus, they say, they chose to make influencing and sharing their lives their job. And while their son is young, "I just don't want him to have a job," Morgan says.

Meghan Moore is a "30 year old Gen Z single mom," according to her TikTok bio. I've been following her for a while, and this is not her first form. When she first started making content, she started videos by saying "Welcome to my miserable pregnant life." After Meghan had her son, Nathan, whom she affectionately calls Natey, she changed her intro to "Welcome to a day in my depressed mom life" as she navigated new motherhood and postpartum depression. And then, when she and her husband separated and divorced, her intro changed again: "Welcome to another day in my single mom life." As of this writing, she has 145,200 TikTok followers.

Meghan started her TikTok journey by making videos about—to put it lightly—not *loving* being pregnant. When we speak, I'm thirty-five weeks pregnant and—to put it lightly—extremely miserable. I have hyperemesis gravidarum, which is a fancy way of saying I've thrown up multiple times a day every single day of my pregnancy. I tell Meghan I really love her early videos because she didn't flinch away from the reality of how miserable pregnancy can be. She congratulates me with genuine excitement. We bond for a moment over both having springtime babies, and it feels good to be included in the motherhood club even before I have my baby.

Meghan's decision to keep her son out of her content started with a Reddit page, as lots of good stories do. When a snark subreddit entirely focused on Meghan was created to dissect (and talk shit about) her every move, she stopped showing her son online to give the haters less ammo. But, she asked herself, was she really going to let Reddit stop her? So she showed Nathan again—until the fake Facebook page popped up, taking her content and pretending to be Nathan's mom. "I was like, 'This is gross. I'm done showing him,'" Meghan remembers. "I was like, 'If I wouldn't let him have social media as a kid, why should he be all over mine for the world to see?' I just stopped showing him, and I don't regret it for a second."

The decision arose out of concern for his future privacy. If Nathan decided as a teenager that he wanted to be on social media, that would be one thing. But as a toddler without the ability to make an informed decision, "I just don't think it's worth exposing him." Meghan felt weird about the ways people were getting attached to her son, and it bothered her that they felt entitled to tell her how to parent him. ("But granted, I was also asking for advice," she says.) Between privacy concerns, Reddit haters, and the weirdness of someone pretending to be Nathan's mom, she was done. Plus, she says, her marriage was falling apart, and the news about Ruby Franke, the mom blogger charged with child abuse and sentenced to prison time, was breaking. "I never want to be a mommy vlogger like that, expose my family. I can expose myself. But doing that to my whole family, I just don't like it. I don't regret posting him. I'll tell him about it one day. I'll have no shame in being like, 'When you were first born, I was so excited to show you. Then I decided to respect your privacy.'"

Six months later, Meghan reversed course, and her son is featured in her videos again. Her content is focused on herself—her videos are about what it's like being a young single mom going through a divorce and showing her days as a hairstylist and taking viewers through her dating misadventures (in one video, a paramour accidentally sent her flowers meant for his ex-wife)—but her son pops up here and there. The focus is not really on him. I wonder if that's why things changed for her. Maybe she's found a way to make her content be about her but let him be a part of it. Maybe there is a sweet spot. Maybe she's found it.

The first TikTok video that went viral for Jillian Kalbaugh, thirty-nine, was one of her baby son sucking up a spaghetti noodle. "It went, like, super mega viral, like millions and millions and millions of views. I was like, 'Oh great. My kid's famous. Let's ride this train.'" She adjusted her content in the hopes of hitting the viral lottery again, posting videos of her son trying different foods like chicken wings and lemons. On top of that content, she made daily vlogs about mom life.

Jillian is matter-of-fact about her ambition to be viral and profit off her content. "I guess I thought we would get millions of followers and start a YouTube channel and become super famous from that," she says. So many parent creators I've spoken to hem and haw—they never meant to go viral, they never expected to become famous, it was never their plan to quit their jobs. It's hard not to be skeptical of that. But Jillian tells me straight up: She wanted to get famous. "It seemed like the thing to do," she says, and I imagine her shrugging on the other side of the phone.

She focused on strategy, posting consistently three times a day and attempting to capture viral content. Sometimes it was fun, but sometimes it was really frustrating, like when she was directing her kids and they weren't going along with what she had planned. "I tried not to be a director, but I had a set piece of content in mind, and I would be deleting and drafting and reshooting it, and it stopped being a natural moment," she says.

Since her first viral video featured her toddler son trying new foods, she focused on making that kind of content, which left her older son on the sidelines. "My older son would hop in the frame and jump in and be like, 'Hey' or 'Hi, everybody.' I was like, 'Sorry, this one isn't about you' and asked him to leave the shot. It was frustrating because I just wanted to get this quick little shot in so I could edit and post it and go about the other stuff. Then [I would] feel bad that my older son wanted to be in the spotlight and have attention. I was unconsciously pushing him aside so I could get the shot." Jillian would explain to her older son that she had to film his younger brother at the moment, but she'd do something with him later. Sometimes, her efforts would pay off, like when a viral video hit right and she got 5,000 new followers in one day. But she never quite took off like she was hoping to, only reaching 40,000 followers at her height. Then a few things happened at once: Jillian got pregnant and she was "sick and miserable" which left her posting less and, the content she did post didn't seem to perform well. "I know a lot of people have said this, but [the videos] just stopped getting views," she says. The morning sickness and crushing fatigue of early pregnancy combined with her lower views made Jillian rethink her plans to become a famous YouTuber.

And then came the death blow: An anonymous TikTok user had liked one of her videos and Jillian found herself looking at the user's profile picture, thinking the kid featured in it looked familiar. Then she realized "the profile picture was a picture of my son from a Christmas card a couple of years ago. It was so scary and gross. It's hard to explain, it was just this feeling in my gut, like 'Oh my god.'" At the same time, the Wren Eleanor scandal, in which TikTokers targeted a mom influencer who regularly posted videos of her toddler daughter that could be considered inappropriate (like the child licking a popsicle) was taking over TikTok. The confluence of these factors helped Jillian decide what to do. "I instantly started removing and privating all of my videos. For me, it was like, that's it. This can't happen anymore before anything gets worse."

Would it have been harder to make that decision if things were really taking off? "Yeah, for sure," she says. Even with the slight success and minor brand deals she had, it was still hard. "I got a lot of stuff the boys wanted. That's how I almost excused it in a way." Now, Jillian has given up on her dreams of social media fame. She says the path she chose may not be right for every family. "Who am I to say what's right and what's wrong?" she wonders. "It's still a gray area."

This conversation about this gray area works best with a nuanced approach, rather than through a set of binary choices (i.e., either you post your kids online and you're a monster bent on exploitation, or you don't and you're a saint forgoing social media fame for the greater good of your child.) We live in a strange, complicated world, and I don't know what the right choice is for parents. It's something I'm struggling with myself as the mother of a young child. I haven't posted her name or face online and only share photos that feature her tiny, dimpled hands or the back of her round head. I've talked about this with my husband a lot and even now, I can't really explain why I don't want her to be online. It was just an instinct that arose when I was pregnant—this instinct to keep her to myself and to our families. She's just so precious to me, and there's something that feels almost like an act of dilution in the act of sharing her with the greater world.

My husband gets this, kind of. He goes along with my boundaries even though, if he were solely in charge, things would be more lax. And I feel bad—I understand him wanting to post photos of her to share with his broader circle of friends and family, but something in me just balks at the idea. Am I being too strident? Probably. How does it affect my infant to have photos of her face online when she's still growing into her features? What exactly do I think I'm protecting her from? Or is it not about her at all; did this decision arise not out of a protective instinct but an anxious one? More and more parents are choosing to share photos of their children only on their Close Friends settings or with emojis pasted over their faces. The parents I spoke with who engaged in this practice told me they were worried about online footprints and how quickly artificial intelligence was advancing. They fretted about how photos of their children could be used or misused and wondered about whether or not their children could actively consent to being shared online. There was also, they told me, an underlying sense of unease. It felt dangerous in some way, to share their kids online. They couldn't say exactly why.

Perhaps there's a level of paranoia in this decision that is difficult to face, but I also think it's the normal pendulum swing of a culture that went from oversharing everything that happened to us to wanting to keep our most precious things private. Just a few years ago, it would have been unthinkable for an influencer or even a regular parent to post their kids only from the back or with emojis over their faces. Now, it's just another choice for parents to make.

On May 12, 2025, Trigg Kiser, three-year-old son of mega-popular mom influencer Emilie Kiser, drowned. Trigg's name was not immediately released to the public but footage of the Kiser family home, where the incident took place, was shown on local Arizona broadcast news. Fans who had spent years watching Emilie (on TikTok, where she has 5.1 million followers, and Instagram, where she has 2 million followers) recognized her house and took to social media to share their theories that the three-year-old boy who had drowned was indeed Trigg.

On TikTok, Emilie was bubbly and approachable, like all the most popular mom influencers. She looked like the girl next door, if the girl next door had achieved viral fame. In her content, she brought viewers through her daily life as she got her kids ready for bed and went grocery shopping and did her makeup. Trigg was a staple in her videos and viewers came to feel like they knew him. More than that, some felt that they loved him. For days, creators made videos comparing Emilie's house to the house shown on the news. They pulled photos from Zillow listings and Instagram posts. As I reported for *Rolling Stone*, there were videos of creators tearfully praying for the safety of Trigg and comment sections in which people sparred over whether or not Emilie had been home at the time of the alleged drowning. Users pointed to the fact that Emilie and her closest friends, a few of whom are also famous influencers, hadn't posted since the day of the reported incident.

On May 18, it was confirmed that Trigg Kiser had died following the drowning incident. Seemingly instantaneously, TikTok users searched government websites, even going so far as to post screenshots from the Maricopa County Office of the Medical Examiner confirming his death. People posted videos of themselves sobbing; there were montages of Trigg created out of Emilie's old content and set to sad music. One video with hundreds of thousands of views offered Emilie and her family a place to stay if they needed to leave Arizona. Another account posed as Emilie posting mournfully for her deceased son. Along with the expressions of grief, people lobbed accusations and blame. It was Emilie's fault, some insisted; no, others said, the fault belonged to her husband, Brady. I saw hundreds of comments insisting they had once seen a video in which Emilie had said a pool fence wasn't aesthetically pleasing. Even though no video has ever been found, the rumor continued to spread. Emilie—and her husband but mostly Emilie—were raked over the internet's coals.

As I scrolled TikTok in the days following Trigg's death, I couldn't get away from these videos. But I saw something else too: mom influencers questioning posting online at all. One mom creator posted a TikTok of herself overlaid with the words "Anyone else questioning how much of their life they share online after seeing Emilie Kiser get ripped apart by the internet?" Ellie, a creator who posts about being a mom in law school, posted a video that said, "I just want you guys to know this will be the last public video from me. The things that I've seen after someone going through the worst thing imaginable are vile and disgusting, and I don't wanna share my life. I don't wanna share my kids." I reached out to Ellie to see if she wanted to speak about this further, but she said she didn't have anything to add.

One mom creator who has hundreds of thousands of followers, whom I'll call Kelly, told me that she was stunned by the lack of sensitivity of viewers and fans following the Kiser family tragedy. "People watch creators on the internet, and they don't realize that they're real people with real lives and real families outside of what they see on the internet," she says. Because of the intensity of the parasocial reaction, Kelly is considering getting off TikTok altogether. "Having a following can be a creative outlet and a monetary blessing for sure, but the cost is your privacy and your peace. No amount of money is worth that. The money you can make from social media is not worth this. Everything that happens to you, good or bad, will be dissected by the internet. Nothing is worth your privacy and your peace."

On August 28, 2025, Emilie Kiser returned to social media. In her posted statement, she wrote, "I now have seen through this tragedy how relationships online lack boundaries, especially in protecting children's privacy. Moving forward, I will be establishing more boundaries with what I share online." After her statement, she went back to posting. It was almost like before, with one notable exception: she never featured her youngest son, Teddy.

Libby Ward, @libbywardofficial with 1.1 million followers on TikTok and 717,000 on Instagram, is the kind of mom influencer who makes it clear to me why people follow mom influencers. One of her pinned TikTok videos starts with "Maybe it's not 'mom rage.' Maybe it's Mom is doing everything for everyone and having less of her needs met than anyone else in the house. All while society tells her to calm down. And shut up. Because this is motherhood and you chose it." When I watch it, I immediately send the video to my sisters and mom friends.

A quick scroll through Libby's content shows nothing of her children. There are mentions of them here and there, enough that I know there are two of them and I think they're somewhere between the ages of five and nine. But other than that, there's nothing else. Libby is a mom influencer who doesn't show her kids online, and she's built an empire out of it.

Libby, thirty-six, lives in Canada and before she became one of the leading voices on motherhood online, she worked as an educational assistant. The kids she worked with told her about TikTok, and she downloaded the app. "I fell in love with how vulnerable people were," she says. It felt different than Instagram, where the presentation style du jour was aspirational and perfect. On TikTok, Libby could be Libby—messy and imperfect and trying.

In the beginning, before she knew TikTok would change her life and propel her career into a different stratosphere, her young kids were featured in her videos. The whole family would do trends or dances—it was a fun, silly way to spend time together. But when Libby's following started ballooning, she second-guessed her decision to have her kids online, and she and her husband chose to take them off totally. "Over time, I really began to understand the layers of being grateful that I made that decision," Libby says. What were those layers? "Their safety, their privacy, their ability to grow up experiencing childhood and life not through the lens of public perception." It's hard enough for Libby as an adult to deal with negative comments online, and she just couldn't imagine subjecting her young children to that. She's recognized in public now too, and she didn't want that for her kids.

Libby found herself thinking of how appearing in her online content would impact her kids as they grew up and built lives of their own and formed relationships. "And I want them to be able to share with people about their life on their terms," she says. "I want them to be able to have a sense of control over what information other people have about them and their lives." A fair amount of Libby's content revolves around the concept of being a "cycle-breaker," which Kathryn Jezer-Morton defined in *The Cut* as "making different choices than one's parents did, thereby breaking cycles of harm that had been passed down through many generations." If that's her focus, Libby tells me, to break cycles and raise children that don't have to recover from their childhood, how can she choose to make them into public figures without their informed consent? But that doesn't mean it's easy. "My kids are my pride and joy, and I absolutely adore them, and I think they're adorable and smart and fantastic and all those things," she says. "But sometimes I feel like I'm limited in being able to share that fully because I don't show them."

There's a part of me that's surprised Libby was able to become as successful as she has without showing her kids. I wonder if there was ever a moment when she thought she was missing out on a brand deal or sponsorship opportunity because of her choices. Libby says that while she can't be positive that she's missed out on deals because of her decision, she thinks it is a possibility. Several times, Libby has been passed over for a sponsorship or brand deal only to later notice that the companies chose "creators that show their children a lot and show their reactions and show their lives. And so you kind of put two and two together."

Even though she doesn't show her kids, there have been moments that, looking back, Libby wishes she hadn't shared. In one instance, her son was going through a phase that Libby worried would lead to him being made fun of in school, and she posted about it, which, ironically, led to him being made fun of online. "All the stuff that I was most afraid of happening, like the whole reason I made the video was sharing my perspective and my fear as a mother, and then to hear things about him just wrecked me," she says. "I was like, 'Oh my gosh.' It just hurt me so deeply. Like, that mama bear comes out."

If the data are correct and so many kids aspire to be influencers, and Libby's kids have a mom who is an influencer, are they dying to get in on the action? Yes, Libby tells me, especially her daughter. "There was a while when she would ask and be like, 'Why?' Like, 'Why can't I?' And so I explained informed consent to her." Consent, Libby said, was her daughter wanting to be in her content. But informed consent was being able to understand the ramifications of being in her content. "And you know, I would just explain, 'At your age, there's no way for you to really understand the long-term or even, like, short-term effects of, like, what it'll do to, like, be in my content. So even though that would be fun for you, my number one job is to protect you. So you might be disappointed.'"

In my conversations, both on and off the record, with mom influencers who have taken their kids offline or never shown them in the first place, there are common threads woven through their decisions. They're hesitant to go on the record with their views on showing kids online, lest they be seen as shading other mom creators. Every mom knows what's best for their family, they tell me. But more than that, they say that the reactions people have had—the insistence that they are entitled to see the influencer's children or the outraged belief that the influencer has taken something from them, the fan, by not showing their child—has done nothing but solidify their decision. See? they tell me. *This* is why.

9

Imaginary Friends:

Model Mothers and the Cure for Loneliness

I am frantic for a model of how to be a mother. I look to my sisters and my friends and my own mother. I read parenting books. I listen to podcasts. But mostly, I look to my phone and the mom influencers who live inside of it.

If you thought that writing this book would desensitize me to the lure of mom influencers and family vloggers, you have more faith in me than I deserve. The pull of mom creators is so strong, and motherhood is so lonely. Why didn't anyone warn me about the crushing loneliness of new motherhood? Or if they did, why didn't I listen? Sometimes I feel like I'm floating on an island way out to sea, and the only things I have are my baby in my arms and my leaking breasts. It's not that I don't have help. I have a wonderful part-time nanny; my mom and mother-in-law take turns visiting and helping out; I have a supportive husband. Even with all of that, I sometimes feel wildly alone. Well—no, that's not right. I'm never alone because I'm always with my baby. But I'm lonely.

Motherhood in America seems to be a uniquely lonely and isolating experience. Not only do we lack the social safety net of basically every other country (seriously, we are one of *seven* developed countries in the world without federally mandated maternity leave), but the hyper-individualism of American culture leaves us siloed in ways people in other cultures aren't. We are American, so we do it alone. We are American, so we pull ourselves up by our bootstraps (despite the fact that this is literally impossible). We are American, so we do it all without any handouts from the government (please sense the sarcasm). We live in single-family households, often far away from loved ones and support, and the overwhelming weight of default parenting still tends to fall disproportionately on mothers.

As I'm writing this, Donald Trump is president for the second time, and the experience of American motherhood has never been more threatened. Pregnant women are dying from being denied abortions. They're bleeding out in hospital parking lots and crossing state lines to access care. Some experts worry that birth control may be under attack next as pronatalist conservatives decry the falling birth rate. Conservative politicians are pushing for more babies in the US while cutting the federal programs that support childcare. At two weeks postpartum, 1 in 4 American women have returned to work. For context: At two weeks postpartum, I was still bleeding and on Percocet from my C-section.

This is not how we were supposed to be mothers, and I can't help but think that these factors are part of the complex interplay of why mom influencers and family vloggers are so popular. As mothers in America, we're missing something—and influencers can help us feel like we're finding it. But are we actually? Or is the sparkling content of influencers just highlighting the void in our own lives?

Ilyse Dobrow DiMarco, a psychologist and author of *Mom Brain*, asks moms in her practice what draws them to mom influencers. Part of the draw, her patients tell her, is the aspirational and instructive nature of the content. Moms tell DiMarco they find tips on how to throw great parties or pack Insta-worthy lunches for their little ones. There's the opportunity to learn skills that will help moms deal with tantrums or decide when to sleep train. "That's in theory," DiMarco says. "In theory, they say it's aspirational and I learn [from them]. But in practice, they look at all this content and then end up feeling badly about themselves and feeling kind of less than and diminished. So it's clearly not working in the way they hope it will." Research backs DiMarco up: one study found that parents who consumed mom influencer content had less confidence in their own parenting skills than parents who didn't follow those accounts.

This makes sense to me instinctually, but I still can't stop scrolling. As I'm writing this, I'm not yet a year postpartum and my phone is in my hand more often than I'd like to admit. The draw of mom influencers is particularly intense in the pregnant and postpartum periods, a 2023 study found. And it really does feel that way. On TikTok, I watch a mom post a time-lapse of her cleaning her kitchen until it sparkles, and I think about my own kitchen, messy and begging to be cleaned, time-lapsed or not. On Instagram, I scroll past perfectly matching and posed toddlers and think about how my daughter recently spent the entire day in pajamas because she was fussy and I was tired. On YouTube, I watch a video where a family bakes cookies together, and I wonder if I should learn to bake so I can do the same with my daughter when she gets older. On every platform, the message is the same: The mothers are perfect, and if they are not perfect, they're perfectly *imperfect*. I leave every platform feeling the same: like a shitty mom who doesn't clean enough or bake enough or try hard enough to look beautiful every day. Mom influencers, mostly, make me feel like shit. I reach for them because of the astounding loneliness of American motherhood, but the image they reflect back to me only makes me feel worse. And yet they are the only people I can reliably find showing what some version of motherhood looks like as I drown in its isolating reality. The rest of the world spins on as I clean spit-up out of my hair and wonder when I last showered.

Sara Louise Petersen, author of *Momfluenced*, tells me mom influencers and family vloggers help us find meaning in our own experiences of motherhood. "Being a mother is so inherently private," she says. "A public mother can be validating. It can be, I think, in a way, a little titillating. I think it's such a vulnerable identity to carry and mothering is such a vulnerable experience. Viewing mothers publicly mothering feels almost like at any point it could veer into emotionally complicated territory. I think especially once you're a mom, you just know in your blood and guts that motherhood isn't all just pretty and isn't all the imagistic shit that we're fed. I do think there's a certain level of: Will we see the other shoe drop with this account? Will we see between any of the cracks? Because it's such a human endeavor, which means it's a messy endeavor."

Petersen has put her finger on something that I've been trying to understand about the fascination with mom influencers and family vloggers. Of all the things I've been surprised by since becoming a mother, the sheer physicality of mothering may be the most shocking. A series of images flash through my mind: my daughter being tugged from my stomach when labor turned into an emergency C-section; the blinding pain that left me dizzy the first time I tried to sit up after giving birth; my husband drying my body after a shower because I couldn't bend down without crying; my daughter's tiny mouth suckling at my engorged and dripping breasts, tender and hot to the touch; the heady exhaustion of those first few months when I watched the world go by through bleary eyes; the scream of my daughter's cries, which left me feeling like I would burn down the world to make them stop; the thrumming of my temples, a migraine constantly threatening to bloom from lack of sleep; and of course, the tiny body that I made, that relies on me entirely, that lives and grows only through my efforts.

This, at least, is my experience of motherhood. It is messy and it is painful and it is not suited to fit inside a square box on Instagram. Petersen is right—becoming a mother has shown me how much of it is blood and guts versus the posed stylized photos and videos we're constantly fed. The difference between the way influencers present motherhood and the actual experience of it is so stark that they don't even feel reflective of the same experience. On Instagram, Campbell Puckett (@campbellhuntpuckett, 1.4 million followers) poses prettily with perfectly done makeup and blown-out hair after giving birth to her daughter. In my real life, I was bleary-eyed from anesthesia when I first met my daughter, and the photos of that moment are something I don't like looking back on. I love Campbell's photos because they show motherhood as I want it to be—beautiful and neat and with mom at the center, smiling serenely while holding a baby. The content is aspirational—and it inspires us.

Emma, a thirty-two-year-old mom, told me her favorite mom influencer is @taylorgiavasis (1.4 million Instagram followers). She loves Taylor's "aesthetic and the way she promotes femininity." Following the account has inspired her to lean into gentle parenting and breastfeeding and even making more food from scratch. It leaves Emma feeling inspired. Scrolling Taylor's account for the first time (how are there *so many* mom influencers?), I can see why: Her version of motherhood is clean and joyous and gentle and shoppable. She is that perfect mix of beautiful and attainable—in the same photo, you can't believe how gorgeous she is, and she also shows off her stretch marks, reminding you she's just like you after all. (Just like you but better.) The you you could be if you gentle-parented and made sourdough and baby food from scratch and used organic wipes (#coteriepartner). "Raising humans, figuring life out" reads Taylor's bio. From where I'm sitting, her version of motherhood seems figured out in a way I can only aspire to.

On Instagram, TikTok, and YouTube, motherhood seems like something that can be perfected, if we only curate our algorithm just right and follow all the gentle parenting influencers and baby-led weaning influencers and exclusively pumping influencers. (Yes, these are all real categories for influencers. Influencers are like porn categories—if you can think of it, there's an influencer that exists for that specific niche.) But even the "relatable" influencers we choose to follow for parenthood tips can leave us feeling as badly as the perfectly curated mom influencers that are on the other side of the coin. Then there are the mom influencers who make us feel like the mess is just a part of motherhood.

Elizabeth, a thirty-year-old mom, told me she started following Val of @lovelyluckylife (178,000 Instagram followers) five years ago when her son was a baby. "It was her seemingly genuine frazzled-ness that was relatable," she says. As I scroll through @lovelyluckylife, I see what Elizabeth is talking about: In a pinned post, Val walks through her messy house and closes the dishwasher with her foot while the text over the video reads "welcome to the side of instagram where our houses look like kids actually live there, our uniform is leggings and a pullover & we laugh about how insane motherhood can be." "It felt a bit like [giving] permission to do things imperfectly when I saw Val talk about it. Need to bribe your kid with M&Ms to potty train? Go ahead! Val did it!" Elizabeth says. But even relatable mom influencers like Val of @lovelyluckylife seem to have motherhood figured out in a way that I don't. In a post hashtagged #walmartpartner where Val stages flowers, a framed photo, and a marble catchall for a gift (complete with a caption including "Comment TEN If You Want My List of 25 #walmartgifts Under $10 & I'll Send Them via DM"), she's poised, creative, and making her work look effortless.

In the vast wilderness of American motherhood, without social support or villages or even acknowledgment, we're left searching. Emily

Hund, author of *The Influencer Industry: The Quest for Authenticity on Social Media*, says the innate loneliness and isolation of motherhood is what draws us to mom influencers. "You are on this journey alone," Hund says, and my stomach hurts from the truth in that small statement. "Even when you have all the support in the world, you are fundamentally alone. When there's someone who maybe is offering help on a subject that you feel like you need help on, or someone who is sharing something like a parenting fail or something that they feel like they messed up about or whatever. That makes you feel better. You can't help it. Being a parent, again, is riddled with such acute uncertainty, too. It's very appealing—that promise of connection and certainty, no matter how fleeting or false it can be."

Even as I follow mom influencers and family vloggers ostensibly for research for this book, I find myself reaching for them when I'm most isolated—like when I'm nap-trapped under my baby who will only nap on a warm body, or when I'm up in the middle of the night, half-awake and breastfeeding. In the glowing rectangle of my phone, I feel my heart clench at an Instagram post in which @annalee15 (386,000 followers) talks about postpartum rage, and then I roll my eyes but secretly feel jealous at a perfect photo of @laurenluyendyk (1.3 million followers) perfectly matching her toddler twins. On TikTok, one of the teen moms I follow, @rocklyn.belle (91,000 followers) just gave birth to her daughter Wrenleigh Bo after days in labor, and @lottie..weaver (2.7 million followers) and her family have moved into a new, obviously much fancier house, with one of the captions thanking her TikTok fans and again, I'm jealous. I scroll and watch @ustheremingtons (972,000 followers) nursing her teething baby while crying, and for a second, I feel less alone. I scroll again and @juliawise4 (1.2 million followers) is talking about why she wakes her baby up from naps, and I feel less crazy about my obsession with wake windows. Another scroll and @homeofizzy (40,200 followers) is blinking into the camera about how she is so tired she could cry, and I feel less tired just knowing someone else has said it out loud.

Amanda Montei, author of *Touched Out*, says followers are looking for a sense of community. "They're looking for some recognition, a feeling that they're not alone, even though we know spending time online just makes us more isolated. They're looking for a way out of that banal drudgery, that boredom that we have with domestic life often. But they're turning to these images that are so aspirational, the promise that maybe it doesn't have to look as messy and dry as it actually is."

When I watch mom influencers perform motherhood online, it seems so much more glamorous and peaceful than my own motherhood. Sometimes, when I'm doing the dishes or filling bottles or folding the laundry, I imagine a time-lapse of the chores being done—a common style of video on these accounts—and it motivates me. In the writing of this book, I've occasionally found myself narrating my day for an imaginary audience like the mom influencers and family vloggers I've studied. *"And then we went for a walk because it's so important to get outside,"* I think as I strap my daughter into my Ergobaby *(#sponsored, if I were a true influencer)* carrier. *"We walked for two miles, and I got my steps in, and E loved looking around the neighborhood. After that, it was time to start our bedtime routine with a bath."* I watch myself from above like some kind of omnipotent god. I imagine a world in which my motherhood is commissionable and perfect and linked, where the work of mothering is valued. In the imagined world, I am not lonely. I have thousands of friends waiting in the screen of my phone.

"I'm your biggest fan," the song—"Anyone Else but You" by the Moldy Peaches—plays over a mash-up video montage of clips of Addie McCracken with her toddler son. Addie is @addie_mccracken123 with 7.5 million TikTok followers; Addie's fiancé is @bradystauffer with 1.9 million followers. "I don't see what anyone could see in anyone else but you," the song continues. The montage has been posted by the user @addyfilmz, a fan account dedicated to Addie and Brady and Rustyn that has 11,500 followers. "A + B + R," reads the bio, "ty for 10k !!"

Addie, a young mom, first went viral making videos with her very cute toddler son Rustyn. "Do you love Mommy?" Addie asks over and over in her second-most popular video, viewed over 59 million times. "No," Rustyn answers as he smiles. It's *really* cute. I remember showing my husband the video when it first went viral and I was pregnant and scrolling endlessly. That was when Addie was just another young mom posting on TikTok—before virality found her and changed her life. Now, as I scroll nine months later, I see her and Rustyn in ads for Google Photos. One of her most recent posts was a sponsored video of "one of my go-to holiday gifts this season!!" for the up-market skin care brand L'Occitane. Addie is pregnant again, and the little family is moving into their new house (which they don't explicitly say was purchased with TikTok money, but it's a guess I'm comfortable making; an expert who is familiar with Addie's brand deals estimates she is making between $2 million and $3 million annually). They're also getting a puppy. They seem happy. Currently, they're one of the most popular families on TikTok—which is what leads me to the fan account @addyfilmz.

The creator behind the Addy fan account is thirteen years old. Over TikTok messages, she tells me, "I made this account to edit them because when I was at my lowest I found videos of them and it really saved me!" What was happening during her lowest time? I ask. "My grandpa and grandma got a divorce so it was really hard for me," she writes, "I find them fun to make." This is her only fan account, she tells me. When I ask what she likes so much about Addie and her family, she doesn't respond.

The top strata of family vloggers have tens of millions of followers and subscribers and literally billions of lifetime views. Much of the content in this corner of the influencer industry is almost wildly mundane. The camera follows families as they make dinner and do laundry and play card games. It's objectively not that interesting—except that, for some reason, people can't stop watching. To figure out why, I posted a Google Form inviting people to tell me about their family vlog fandom. I got over two hundred responses in just a couple of days. I asked fans to tell me why they were drawn to family vloggers and I realized quickly that there was a common theme within the hundreds of answers. It broke my heart a little. Over and over, fans told me they were trying to escape their own less than ideal home life and the presentation of perfect family life from vloggers was how they achieved that escape.

Consider these responses:

> My home life wasn't particularly great and I struggled to connect with those around me but when I watched these videos, through my twelve-year-old eyes I saw a big happy family.

> I was a young, depressed, lonely, financially poor child. Watching the [redacted] & their family made me so happy because for a little bit, I could escape my terrible home life & see how other children were enjoying their life. They had it all at one point, money, friends, family, materialistic items, everything an 11 year old could wish for. I watched them every day after school when they did daily vlogging.

> I watched them during a time in my life when I had very little family around me. One of my parents had just passed away,

and it was just me and one parent left at home. Their content appealed to me because I really felt like "a part" of their family. They always had so much fun, and seemed to live the perfect life.

And once they started watching, the fans and former fans told me, they got hooked quickly. "I watched them almost every time they uploaded," was a common answer when I asked how often they tuned in. "I watched every video" was another one. Soon, the viewers said, they felt like the distance between themselves and the families they were watching online shrunk.

> it felt like I was part of a community and they were like my cousins in a way

> I felt like I was part of their family

> i always dreamt that i was a part of their family. I watched them every day

There were moments that felt too vulnerable to be on camera, the viewers told me. Among their stated examples: the video in which the dad of a family with millions of subscribers on YouTube asked his teenage daughter if she was flirting with her crush at school. The daughter, seeming embarrassed, demurred; the video had millions of views. (That same dad wrote a book with his then nine-year-old son called *Fat Dad, Fat Kid* that professed to explore the question "Does fat dad have to mean fat kid?") There was the family that regularly had puberty talks with their children on camera, despite how uncomfortable the kids looked. There was the vlog in which a kid said she had to be homeschooled because a stalker had found her school—and the fact that the parents didn't change what kind of content they uploaded online after that. There was the young teen shaving her legs for the first time, the act caught on camera and uploaded to YouTube. All the videos of kids throwing up or crying or having tantrums. Endless videos showing milestones in the lives of family vlogger kids: their first crush, their first breakup, their first period, their first sex talk, their first time going bra shopping.

The deeply personal nature of the content they were watching made them uncomfortable, fan after fan told me, but they couldn't stop themselves from watching. The content just kept pulling them in. "Some part of [the fascination] is just truly wanting to know other people are doing it," explains Jess Grose, a *New York Times* opinion writer. "Remember when you were a kid and you would go over to someone's house for a sleepover? And there was just this whole other world of how families were functioning different from your own. Part of it is just natural human curiosity about how other families live their lives." One former fan I interviewed echoed Grose's reasoning, telling me they used to watch Ellie and Jared, a channel with 1.6 million YouTube subscribers, religiously because "it was a little view into a completely different life." One respondent, who obsessively watched family vlogs between the ages of ten and twelve, told me that she was especially drawn to the content as an "isolated homeschooler. So I sort of lived vicariously through the vlogs. It was my way of experiencing the world when I was stuck at home all day." She watched the videos every day, staying on top of the upload schedule to see what her favorite families would do next.

Grose tells me in her research into this content that she's found that people are often drawn to creators who are raising their families in a way the person viewing the content did not experience. It can be healing to follow a gentle parenting influencer who is constantly apologizing to her children and validating their feelings when you didn't grow up in a house where that was the norm. "People are looking for answers very desperately," Grose said. "It really is much harder when you don't have models for how to parent in a way that you felt was good." And influencers can stand as those models, as you watch them parent their children in a way that feels both astronomically far from how you were raised and, somehow, possible for you to emulate as a parent. As I scroll through TikTok and YouTube, I see family vloggers practicing #gentleparenting, a popular parenting style (at least online) that strives to approach a child's every action with empathy and understanding. They apologize to their children when they make mistakes; they ask them their preferences. Despite myself, I mentally take notes for my own parenting.

There is also part of family vlogger fandom that can simply be chalked up to rubbernecking. It's the same way we crane our necks to look at car crashes. We are all drawn to human spectacles—and I'm not immune to this. One of the family vloggers I'm most drawn to is Karissa Collins, a mom of eleven (!) children. All her children have *A* names—Anissa, Andrae, Annistan, Anjalie, Andersyn, Aynjel (pronounced "angel"), Ansyr (pronounced "answer"), Anchor, Anthym (pronounced "anthem"), Armor, and Arrow. She's religious to the point that she considers IVF evil and any form of contraception—including pulling out or timing your cycle to not have sex during your most fertile periods—ungodly. One of Karissa's recent Instagram posts is a sweet video of her kids playing, accompanied by the caption "Believers I beg you to wake up. We're sacrificing our future children so we might live. It's in direct rebellion to the word of God. #abortionismurder #family #children are a blessing." Karissa had 1.6 million Instagram followers before her account was taken down because of what she claimed was censorship of her views. On her new account, she has 622,000 followers.

I originally followed Karissa when I became interested in interviewing her for this book. She agreed to an interview, and we set up a time, but the morning it was supposed to happen, she canceled. Part of why I still follow her is so that I can slide up on her Instagram stories occasionally and try to reconvince her to talk to me. But the other part is that I just can't look away from her life. It is pure rubbernecking and I can't stop. She just lives such a fundamentally different life than I do—she's a mother of eleven children who doesn't believe in birth control or vaccines and does believe in homeschooling and spanking.

In a post from her now-defunct blog, Karissa muses that she always knew she would spank her children, but now that she has a child—who was then one year old—she finds it strange to discipline the toddler for hitting someone by hitting her in return. Plus, she continues, time-out seems to be more effective. Still, she worries. "I don't know if it is just my husband, but he seems to not know his strength and I feel that he spanks too hard, and I don't want to cross the line of abuse. It seems to be a thin line."

It's natural, I think, to be drawn to depictions of family life. Sometimes, we wish our families were like that; other times, we can't help but judge. We wonder why family vlogging is so popular, but the answer seems to be deceptively simple: We want to see how other families function and measure them against ours. "I liked seeing how other people parent their children. I liked seeing how both the parents and the children manage their time and their money," one fan told me, and I think about @lottie..weaver (2.7 million TikTok followers), whose content I can't stop watching. The mom does her three daughters' hair each morning and talks about the importance of putting yourself together, and I'm rapt. They go to the gas station and let the kids pick out sodas and candy. She packs their lunches (that she proudly stocks with snack food because, she says, she's just trying to pack food that they'll eat), and for some reason, I'm still watching. Scratch that. I know why I'm still watching: because I want to see how other people live and compare it to the way I live. I want to clutch my pearls as the kids pour themselves sodas and play on their phones. If I'm honest, I think there's a part of me—of all of us—that wants to watch these families only to be able to hold them up against our own. *Look*, we want to say. *Look how much better they are than my own parents* or, in some cases, *Look how much better I'm doing than they are*. Watching other families comforts us or inspires us or simply functions as a spectacle. A family unit is almost unbearably intimate. No wonder we all have our noses pressed against their windows.

10

From Love to Hate:

Former Fans, Parasocial Relationships, and the Rise of Snark Communities

Alexa (a pseudonym), a nineteen-year-old student who lives in Norway, goes to school during the day, and works evenings at a local store. In between that, she's an unpaid moderator of the subreddit r/FisherFamilySnark. The subreddit, which has around 5,100 members, is dedicated to snarking on the lives and content of the Fisher family (4.67 million subscribers on YouTube): namely mom @madisonbontempo (1.9 million Instagram followers) and dad @kylerstevenfisher (458,000 Instagram followers), who are the parents of kids @taytumandoakley (the account that is home to twins Taytum and Oakley, which has 3.1 million followers and contact information for a talent agent in its bio), @halston.blake (the account of Halston Blake, who has a comparatively paltry 522,000 followers), and @oliverandcohen (the shared account of brothers Oliver and Cohen, who have 163,000 Instagram followers).

Like many "snarkers," Alexa began as a fan of the family. She started watching the Fisher family in 2015, when she was nine years old. She eagerly awaited each new video, like "Birth Video of Baby Twins Taytum and Oakley (Emotional)," which garnered 5.5 million views, and "Hilarious Family Workout Challenge!," with 3.6 million views. "They seemed like this perfect family," Alexa tells me over a WhatsApp call. "And I had a rough upbringing, and I remember watching them because they were comforting. It was just nice to see a happy family, kind of." Four years into her fandom, she stumbled upon the r/FisherFamilySnark subreddit, which bills itself as "a place to snark on the FISH FAM. We are NOT a place to worship, support, or obsess over them." When Alexa first saw the snarky content, which painted the Fisher parents as exploitative and the Fisher kids as exploited, she was in denial. She figured the snarkers were haters who were twisting small moments into things they weren't. But her view on the family Alexa used to love began to change as time went on. Part of it, she says, was that she was getting older and viewing the content differently; she began to worry about the privacy implications of the Fisher kids being the forefront of their content. She also frequented the r/LaBrantFamSnark subreddit, where posters claimed (seemingly without proof) that the majority of Everleigh's followers were grown men, despite the fact that the influencer herself was a child. "And that's when I just started thinking that maybe it's not so great that their whole lives are on the internet for everyone to see because you can't really regulate who sees them," she said. "And it was just something I was very passionate about. Like, [the Fishers] had been such a big part of my life. And when I realized that they aren't that great, I thought others should also know." Within months, Alexa had morphed from an ardent fan of the Fisher family to a moderator on a subreddit dedicated to snarking on every detail of their lives.

As viewers and advocates grapple with the ethical considerations of the monetized childhoods of influencer children, the backlash against family vloggers and mom influencers has only grown louder. And amid that backlash, snark subreddits are flourishing. In these communities, former fans have found a place to continue discussing the creators they used to love. But is snark just another way of being a fan? Is anti-fandom the new fandom?

Let's begin with the word *snark*. Is snark just a cutesy way of labeling something much darker—hating at best and cyberbullying at worst? Amanda Montell, author of *Cultish: The Language of Fanaticism*, tells me the word *snark* was coined by author Lewis Carroll in the nineteenth century to describe an imaginary animal, and then the neologism evolved to mean someone who finds fault easily. "When a label becomes a genre, it provides a permission structure for a certain category of behavior," Montell says. "Snarking became shorthand for something excusable." I really do think the cutesy label snark rebrands behavior we would otherwise find repugnant. The people participating on snark subreddits don't see themselves as haters, bullies, or creeps. They're just snarking! It's not that big of a deal, right?

The concept of snark did not originate with Reddit. In the early 2000s, when the first mom bloggers were exploding onto the scene, there was a website called Get Off My Internets—GOMI for short—which was basically a tabloid-esque forum to talk shit about your favorite bloggers and online personalities. (Back then, fans didn't use the word *influencer* to describe these people.) Between GOMI and Perez Hilton's eponymous blog that tracked and obsessed over the mundane details of celebrities' lives, "the internet really began to change who was participating in this conversation," says Jess Rauchberg, PhD, a professor at Seton Hall University in the College of Human Development, Culture, and Media. "So it's not *US Weekly* or *People* magazine, where there are certain forms of verification that need to happen in terms of how that news or the gossip is reported." Suddenly, anyone with a computer could spread gossip—and they could do it quickly, without the fact-checking protocols of more traditional publications. When influencer culture really started to take off in the mid-2010s, Reddit became *the* place for snark. The subreddit r/blogsnark, which has 197,000 members and brands itself as the place to "snark on your favorite bloggers, influencers, and everything else on the internet!," was founded in 2015, and since then, the world of snark subreddits has blossomed—especially when the object of the snark is family vloggers or mom influencers. There's r/doughertydozen (36,000 members dedicated to snarking on the Dougherty family, who have 1.6 million YouTube subscribers); r/McKnightFamSnark (25,000 members who snark on Brooklyn and Bailey McKnight, twin influencers who have 7.3 million YouTube subscribers, and their family); and r/8passengersnark (67,000 members who gather to discuss Ruby Franke and the fallout of the 8 Passengers YouTube channel, which had 2 million subscribers at its peak).

On r/GDSnark, about 2,600 people tune in to snark about Bethanie Johnson (formerly Garcia), the mom influencer we talked about earlier who has more than 338,000 Instagram followers. When I stumble across the snark subreddit dedicated to Bethanie, I can't stop scrolling. There are detailed logs of her follower count, which seems to be dropping, with exaltations from anonymous accounts that they "love this for her." There are screenshots of an innocuous selfie, with comments making fun of her hair and what one anonymous user suspects are lip injections gone awry. There is a link to a website where snarkers can watch Bethanie's stories and videos anonymously, meaning they don't contribute to her views and metrics, an influencer's lifeblood.

When I connect with Bethanie about the snark page, she's glad to talk. The timing is fortuitous: She just received a text from an unknown number with screenshots from the snark page where strangers were making fun of her. She found that the phone number was generated from a website. "It's bizarre," she tells me, as she was arranging to speak to me about the snark page, "these texts popped up." She muses about who the text could possibly be from and hopes it's from someone she knows in real life. The alternative is that she was doxxed and some anonymous online hater found her phone number. "It's scary that people in real life could be behind it or scary that strangers on the internet could have my phone number. Just so weird all the way around. Either way, it sucks."

When Bethanie first became aware of the community dedicated to snarking on her, she says she had a mental breakdown. "My stomach sank into my vagina literally," she says. "I couldn't get up for a second." She was at dinner with her ex-husband when he told her about the subreddit, and she locked herself in the restaurant's bathroom stall to cry. She felt sad, misunderstood, and helpless, but after spending two days going through every post, she promised herself she would never look at it again. And she didn't, until official news of her split with her husband sent the snark community into overdrive.

Though she never directly sought out the snark, friends and followers send her screenshots and links to the page, asking if she's seen the latest rumor about her. It's difficult, Bethanie says, to face all the lies that are spread about her, but there's just enough truth sprinkled in to make the lies that much more believable. Bethanie doesn't get why people are so invested in her. Sometimes, she pictures the people in her snark community—1,700 as I'm writing this—standing in a room with the express purpose of hating her. If they don't like her, why don't they just unfollow? "I can't really rationalize why people are so invested or care so much," she says. "Like, why do you care?"

On r/GDSnark, there are breathless posts and comments dissecting theories of why Bethanie and her ex-husband got divorced. Part of Bethanie wants to clear it up—to explain why she's getting divorced (she tells me off the record, and I wonder if knowing the truth would make the snarkers quiet down or get louder), but a bigger part of her will never do that because she wants to protect her kids from the knowledge of the inner workings of their parents' failed marriage. And she wants to protect herself. Just because she's made a living sharing her life online, does that mean the public is owed every detail about it? How has the internet facilitated the parasocial relationships we develop with influencers to the point that fans think it's their right to know why a marriage between two people they don't know ended?

The entire career of an influencer is based on stoking the parasociality of the relationship between their viewers and themselves. "A parasocial relationship typically implies that there is a power dynamic and a one-way relationship," says Dr. Brandon Harris, an assistant professor of social media analytics and production at the University of Alabama. He explains that the typical parasocial relationship is between a famous person and their fans. Though fame is not new, the power of parasocial relationships has been supercharged by the internet and the attendant access it confers to famous people. "Shirley Temple did not have her own phone," Dr. Harris says. "The audience did not have direct access to Shirley Temple in the ways that people have access to influencers now. There was no Cameo to go and talk to them. There was no Twitter to follow what they thought at every single moment. There were no Twitch streams." As the amount of information fans can find out about their favorite celebrity increases, so have the rates of celebrity worship, a 2021 study found. With social media, the level of access people have to the people they're fans of has exploded—and with that, so has the intensity of our parasocial relationships.

But what happens when those parasocial relationships sour? "When an influencer that you've subscribed to for so many years 'wrongs you,' all you can do is unleash into a void. Because going through a parasocial breakup is so anticlimactic. You have to show that you are here and that you've been here," says Montell. This hits on something I've been trying to figure out about snark communities: Why are they so densely populated by former fans? Montell's explanation makes sense to me: A parasocial relationship breakup is inherently lonely. It's a one-sided relationship that's ending. So maybe it doesn't have to end. I understand the instinct to turn your fandom into snark. There's something about the metamorphosis that makes it feel like you didn't waste all those years you spent as a fan. There's somewhere to put your old love.

"Hey, what are ya'll doin'?" says @kaitlyn.a.schultz (546,000 followers) in a TikTok video as she perches on a stool in a cropped gray T-shirt and short shorts, her hair pulled back with one of those influencer headbands and a makeup brush punctuating each of her words as she twirls it around. "Nothin'? Cool. Okay I hate to break this to some of ya'll, but if your whole life revolves around watching one person's page, going through all their comments, leaving hate comments, screenshotting their shit, posting it on a different site, talking shit about them there, just being obsessed—I hate to break it to y'all. Y'all are not haters, y'all are fans."

A blue circle pulses around her profile picture, indicating a current story. When I click, Kaitlyn and her two sons skip toward the front door of their house with the following words overlaying the video: "On our way to hit Starbucks, [chick-fil-a], and the target toy aisle with the $$$ we make from ppl who don't like mommy ;)."

If you're spending your time tracking someone's posts, screenshotting their videos, and watching their lives before posting about it all on a community dedicated to them, as Kaitlyn Schultz theorizes, aren't you really a fan? Sometimes, the snarkers seem more dedicated to the subject of their ire than self-proclaimed fans do. But maybe it's not fandom after all—maybe it's "anti-fandom," as defined by Jonathan Gray of University of Wisconsin–Madison: "Opposed and yet in some ways similar to the fan is the anti-fan: he or she who actively and vocally dislikes or hates a program, genre, or personality." Snarkers are the perfect example of the anti-fan: They are drawn together not because of their love of a subject but because of their hate. And that shared hate can build community in much the same way that shared fandom does. "It's like any community online [where] people participate in it to create affinity within the group," said Kathryn Jezer-Morton, a columnist at *The Cut.* "The group identifies itself very strongly and there are ways that people can reliably get a lot of engagement and traction within the group. It's like you're rewarded for being really mean. If we use a sociological lens, we can excuse them and be like, 'Well, they're just trying to make friends on the internet.'"

Caitlin (a pseudonym), a thirty-eight-year-old mother and part-time moderator of the r/LaBrantFamSnark subreddit, was first introduced to the LaBrant family (12.8 million YouTube subscribers) through her daughter, who was a fan of the family. Then the LaBrants posted a YouTube video titled "She Got Diagnosed with Cancer," the thumbnail of which was a photo of their young daughter, Posie. The implication, detractors felt, was clear: Posie had been diagnosed with cancer. Just minutes into the video, Posie is declared healthy and the video turns into a documentary on another child's cancer journey. "What they insinuated by the clickbait title was insane to me," Caitlin says. "I wanted to see if other people felt the same way I did. And I started looking, and I found the Reddit community. I dabble in different conversations on Reddit about all different kinds of things. So this is just another niche interest." Now, Caitlin is part of a group text with her fellow moderators, many of whom are also mothers. She's even met some of them in person. "We're a pretty tight group," she says. "It sounds silly, but it's not so much about the LaBrants anymore. We've really gotten to know each other as humans." On Caitlin's daughter's twelfth birthday, she went to Disney World with another mod and her young daughter.

On r/LaBrantFamSnark, 45,000 members gather to dissect the LaBrant family's every move. There are nicknames (mom Savannah is sometimes referred to as "Satannah") and rules, including a ban on snarking on the children. Caitlin is convinced that the LaBrant family regularly checks their dedicated snark subreddit (though an immediate LaBrant family member told me this isn't true). Does she feel bad about it? "There's consequences to all actions," she says. "When anybody puts themselves out there for public consumption, they're making themselves available to judgment. And some of that judgment is going to be really positive, and some of it's not." The sub has even broken news before, like the name of the LaBrant's fourth child, Zealand, which was posted on the sub before the family announced it. Someone close to the family messaged the mod team and told them the name, Caitlin tells me. They verified the source before posting the name reveal, but even so, Caitlin couldn't help but feel a little guilty for spoiling the name. Her justification? That the LaBrants and other vlogging families sometimes stretch name reveals into several pieces of monetized content, like multiple YouTube videos. "They want to make as much money as possible," Caitlin says. "And I think what it comes down to is none of us feel they should be making their living on the backs of their kids." Viewed that way, spoiling the name reveal didn't seem wrong—it almost felt righteous.

This is what separates snarkers from haters—snarkers have embarked on what they see as a moral crusade. When you read through snark pages, the posters are adamant that they're not just hating—they're raising awareness about whatever issue they think the influencer they're talking shit about exemplifies. In Bethanie's case, for example, they'll say they're raising awareness about moms who seemingly exploit their children or influencers who post too many ads. There's a moral component that I find fascinating, an attempt to intellectualize what could otherwise be considered simply badmouthing. "Their position is that their critique is one of integrity," Rauchberg says. "[Like], 'oh well, this person doesn't disclose their ads or we know that they don't pay their interns or they steal from other creators.' I think snarkers are constantly negotiating in these moments where it's like, 'Wait, are we actually being mean? Or are we not?' I think it's a constant negotiation for them."

If you've never been on a Reddit snark page, well, first of all: congratulations on being a well-adjusted person. But second: There tends to be this idealized vision of snarking, where, Rauchberg explains, "We're not gossiping, we're not shit-talking, but we are doing a secret third thing." One of the ways snarkers define their role is by instituting rules around their snark. On the r/parentsnark subreddit ("your village for snark on the world of online parenting," the sub description boasts to its 22,000 members), there are nine stated rules including "don't snark on the appearances of children and don't body shame adults" and "do not go 'real life' and contact influencers or posters from other social media sites." The existence of these boundaries is fascinating to me—to put up a fence in a community that is literally dedicated to talking shit about people. But it makes sense to Rauchberg.

"It offers nuance and a complexity of what it means to be a snarker," she said. "I think that while snark rejects that, embraces the art of hating, so to speak, and really thinking about, 'Well, we're not necessarily gossiping to gossip, but we're offering critiques about an industry where there's issues about representation, credibility, and authenticity, and we need to address those issues.' But I think those boundaries also show the externalization of navigating those boundaries for not just the snark community, but individual snarkers, and thinking about, well, *When am I crossing that line?*"

Sometimes, snark communities even splinter over rules and other, less-policed snark communities are formed where fewer topics are off-limits, but often there's an imagined moral superiority with snarkers. They alone know what the influencer is doing wrong—and they alone can tell the world about it. In response to a question about what made them join a snark community, one r/LaBrantFamSnark member told me, "We are against exploiting children." An r/duggarsnark member said, "I joined because these people are exploitative of themselves and their families. I like calling out bad behavior."

Psychologically, Montell sees snark communities as addictive. "They are never going to get real answers to their speculation and they don't even want them, I don't think," she says. "At the end of the day, it's way more about how it feels to be a part of these communities, and it's addictive. Every time you check a forum, your body is hoping for a hit of dopamine in the way that every time you scroll through your Instagram feed it's like pulling a slot machine at a casino like it has your brain chemistry in a chokehold. You want to see that other people are upvoting your comment and you will be rewarded for that and there are obviously these dopamine incentives to sharing negative—even or especially—false information that will generate more votes and likes." I'm struck here by the parallel between the allegedly likes-obsessed influencers the snarkers are obsessed with and their own pursuit of a psychological high through upvotes.

There's also a lopsided sense of power. Scrolling through r/GDsnark, I see screenshots of Bethanie's content with posters swearing that she only posted that specific thing in response to something they themselves had posted in the snark community. "I think that a lot of the communities almost want the subject of their snark to perceive them," Montell says. "They want them to lurk in these threads. They want them to see what they're saying."

According to Bethanie, there can be an instinct to address the rumors and lies spread about you as an influencer, but she tries not to. "I think it would give them some fulfillment that I'm acknowledging them," she said. "But, yeah, obviously there's a part deep inside of me that's, like, I wish I could just get on my story or on live and, like, annihilate them. But what good would that do besides a momentary feeling of satisfaction? The next day they're going to be on one again."

When Bethanie started dating someone new after her separation from her husband, the snark page was all over it. They found out the identity of her boyfriend and posted his college records and pictures from his high school soccer team. When she moved into a new house, the amount she paid for it was posted on the subreddit. It all just kind of freaks her out, and it's even made her have fleeting moments of wanting to stop being an influencer altogether. But how else could she make $500,000 a year? "I would love to just be able to live my life and not have to worry about me having a boyfriend offending a lot of people and making them irrationally angry, or me getting extensions [put] back in making a lot of people angry," she said. "But in a perfect world, if I could make this much money not having a college degree at another job, I would be on that in an instant. It's definitely not worth it. I definitely would give it all up to not ever be on the internet again. It's definitely not worth it. But at this point, there's not really anything else I can do."

I hear her. What is she supposed to do, as a woman who started having kids at eighteen and is now a thirty-year-old mother of five with only a high school degree? But she keeps saying it's not worth it, and I think the hard truth is that it *is* worth it. The hate and the snark and the tracking of her every step—it is worth $500,000 a year. Or she wouldn't be doing it. Right?

When I ask Bekah Martinez, a mom influencer with 863,000 Instagram followers, about being the subject of Reddit snark (Bekah is mentioned on r/thebachelor as a former contestant of the show and doesn't have her own dedicated snark page), she says "it legitimately puts me in a state of fight or flight to talk about some of the things that have been said on there." I quickly search through the subreddit and see what she means: There are posts about viewers calling child protective services on her and a slideshow of her "problematic pics" dating back to 2014 when she was nineteen. "I have gone through the murder of a close relative and talking about that doesn't come close to the nervous system overload that happens when I talk about some of my 'cancellations,'" Bekah tells me. "I have always been blown away and actually felt sick to my stomach at how vile people on there can be."

Looking at the snark pages is something that many influencers do in the beginning of their careers, Bekah tells me. It's just human nature. "If you knew thousands of people were talking about you behind your back and you could take a peek at what they're saying—it takes enormous control not to," she says. I know I wouldn't have that kind of control.

Before becoming an influencer, Bekah was more trusting of other people. She gave people the benefit of the doubt and assumed the same grace she offered people would be offered back to her. Now, she knows the truth: "People are waiting in the wings for any opportunity to twist you into something that makes you feel ashamed and embarrassed." As a result, Bekah has changed her content. She's less vulnerable online now than she used to be—though she thinks that may be a good thing. "I save those [parts of myself] for the people who really care about me and who *I* care about."

Bekah isn't the only influencer who finds it difficult to discuss her snark page. When I ask @annalee15 (391,000 Instagram followers, 570,000 TikTok followers) about the posts about her on r/parentsnark, she texts me back: "Ahh I would love to help but Reddit is truly such a dark place for me and fucks me Up so bad, I honestly don't even know if I can talk about it 😅." That's what makes it so important, I counter—people need to understand what it's like to be the focus of snark subreddits. After I make my argument, she doesn't text me back. I don't blame her.

Abha Ahad, a journalist who writes the Substack *Girl Online*, argues that snark communities aren't entirely problematic in the post "in defense of snark subreddits" which wonders if snark subreddits aren't a netizen's way of "rectifying the internet's power imbalance." Modern-day snark subreddits aren't the first of their kind, Ahad says, but actually descend from earlier internet sites like Get Off My Internets (also known as GOMI, which had pages dedicated to the original mom bloggers) and Guru Gossip. "While these forums were vicious with little to no boundaries and rules, snark subreddits have better guidelines in place," Ahad writes. "For instance, r/blogsnark has rules that specify that the snarkers can't reveal information posted on a private or deleted account, not to mock how the subject of discussion is processing their grief or their health conditions, and about posting gossip without proof. Other subreddits dedicated to a specific influencer have guidelines that mandate that the users don't make direct contact with the influencer or body shame them. The moderators also ban users who make homophobic, racist, or sexist comments. They also ensure the influencers' boundaries are respected. For example; if a creator doesn't share pictures of their kids or family online, the snarkers also don't."

Ahad argues that there is a place for snark subreddits in the online ecosystem. The influencer, by nature of their job, makes money by selling viewers an "at worst fabricated and at best curated" version of their lives and though this happens on social media, there's nothing social about it, Ahad says. It's a one-way mirror, where we watch the influencers watching themselves. "At the end of the day, influencers and snarkers are two sides of the same coin. Snarkers are anti-fans, just like the hate-watchers. The fame of the influencer is directly proportional to the popularity of their snark subreddit and snarkers are often as obsessed with their subjects as much as their loyal followers. But they have different functions to perform. While followers fuel influencer culture and strengthen the creator economy, snarkers—with a little bit of personal discretion and better moderation from Reddit—can hold influencers accountable and to an extent, balance the internet's power asymmetry."

Though I find this to be a somewhat starry view of snark pages and what they represent on the internet, I see what Ahad means. The snark page is basically a democratized tabloid for influencers who are the internet's celebrities. Can influencers really expect not to get pushback? Isn't that kind of what they signed up for? And to be sure, there have been moments when snark pages have made actual social change, like when r/thebachelor—a subreddit dedicated to *The Bachelor* franchise that leans into snark—surfaced photos of a contestant on the first Black Bachelor's season partying on a plantation. Rachael Kirkconnell, the contestant, still won the season and dated Bachelor Matt James for years, but the incident spurred a reckoning over the franchise's handling of race in which longtime host Chris Harrison was fired. There was also the instance of white lesbian influencers Lunden and Olivia (who have a joint TikTok account with 895,000 followers). The weekend of their highly publicized wedding, hundreds of their anti-Black and racist tweets were leaked on r/lundenandoliviasnark, casting a pall over at least the honeymoon, from which Lunden and Olivia made sorrowful apology videos before taking a break from their TikTok account.

Snarkers also seem to have had a hand in deplatforming influencer Liv Schmidt from TikTok. "Across various snark sub threads, members explained that they rely on TikTok's reporting tool to mass-flag Schmidt's account for its promotional proliferation of pro-anorexia (pro-ana) content because it violated TikTok's disordered eating and body image guidelines," wrote Ivy Griffith for *Distractify*. So snark can be—if not constructive—at least effective in some way. Racist photos and tweets are unearthed, and an influencer who promotes eating disorder–adjacent content is taken off a platform. The internet, for a moment, feels less toxic.

But do these victories—which are few and far between—justify the sheer volume of vitriol these online spaces generate? They are forums for negativity and offer negative interpretations of every situation. On r/LaBrantFamSnark, users dissect the text eleven-year-old Everleigh LaBrant sent her stepdad Cole on the day he adopted her. "Congrats? You didn't save a kid from the system. Literally nothing about your lives changes," one comment reads. On r/GDSnark, a poster accuses Bethanie of using her boyfriend as "rage-bait," and the top commenter agrees: "Seriously—has 5 kids and has time to do this embarrassing shit." The second comment reads "He grabs her ass as if she has one." On r/kaitlynschultzsnarkk, one post includes a screenshot from one of Kaitlyn's recent videos and an image of a pug with the caption "swipe to see the resemblance."

Influencers are the celebrities of the twenty-first century, but most of the time, their exploits are not covered in tabloids. Instead, they're breathlessly discussed in snark subreddits and comment sections and reaction videos. Snarkers watch influencers fold their laundry and parent their children and celebrate their triumphs, and the parasocial relationship grows only deeper. There's a frustration there in this one-sided relationship, but on a snark subreddit, there is proof of the relationship everywhere. The time and effort and love we've put into the influencer is validated by thousands of other people who are just as affected by them as we are. You don't have to call it fandom or hatred. You can call it snark.

11

Baby's First Sponsorship:

The Economics of Influence

Before **Alexandra Sabol** went viral on TikTok, she lived with her three children in Section 8 housing. Sometimes, she relied on the food bank to buy pantry and household staples to feed her family. In her first viral video, which has now been viewed over 92 million times, Alexandra pours huge glugs of cooking oil into a stained and scratched pan. She mashes meat into hamburger patties and sets them on a different, similarly scratched and stained pan. Back to the first pan: she dumps fries into the oil. Back to the second pan: she flips the burgers and adds seasoning and when she flips them, they appear burnt. She puts them on a plate and starts over with more patties. Her glasses are fogged over; at one point, she pauses to show the camera a cut on her finger. And if you're wondering why a video this seemingly innocuous would go viral, she has a theory. "Because I'm a big person, and people don't like fat people," she says simply. Alexandra estimates she made $4,500 off the video. As of this writing, she has 845,000 followers on TikTok.

Since going viral on TikTok and gaining hundreds of thousands of followers, she's been able to buy a house for herself and her children and return to college. At her last job before TikTok fame, Alexandra made "$25,000 a year maybe." In the first seven months that her content became monetized on TikTok, she made $40,000, and she's hoping to break $70,000 by the year mark. As a twenty-seven-year-old mom who had her first child at eighteen and had to drop out of college because she was pregnant, Alexandra never thought she'd earn this much money. "After I bought my house, I just sat in my house and I looked around and I was like, I never imagined actually being here, being the homeowner, being able to take care of my babies by myself, being able to support them and feed them."

Food is the focus of almost every single one of her videos. Reading through the comments on Alexandra's videos, it seems clear that many people who watch her are hate-watching (not that it matters—a view is a view). The comments range from rude ("I wouldnt even feed this to my dog") to downright cruel ("if your so lazy and tired why'd you even have kids?"). "Cook Breakfast with Me as a Single Mom of 3" reads the text laid across her second-most popular video, which has 38.5 million views, as she pours cooking oil into a pan and then drops in frozen donuts. "Plate Dinner for My 3 Kiddos with Me," is her third-most popular video (29.1 million views), and it features Alexandra unboxing fast food from Wendy's and putting it on plates for her kids. The top comment, referencing the apple slices she tosses onto a plate, reads "the apples are just there for the aesthetic [at this point]" and has 145,900 likes. Most of the comments are mean—they attack the food Alexandra feeds her children, the way she cooks it, and the way she looks while she cooks it. Whereas other mom influencers and family vloggers seem to have genuine fans who take pleasure in their content, almost all of Alexandra's comments are criticizing her. It all makes me wonder: Is her content elaborate rage-bait?

I edge toward this question carefully, first investigating which of her content performs best. "If I post that I got my kids McDonald's or Wendy's for lunch, then that's going to go super viral," she says. "People pay attention to stuff they don't like, to the stuff that irritates them and upsets them." Right, I say. And if I were her, I would have a hard time not leaning into that and posting more of what makes people angry—because making people angry makes them interact with your video, and those interactions equal more money. Is she ever tempted to do that? "I feel like I definitely have increased how much I eat out now that not only do I have money for it, but it's convenient because I'm so busy throughout the day, and also I make good money from it," she says. "I still try to cook, but it is definitely more tempting to go for fast food because it's going to make more money. Plus it's easier for me as a busy mom."

Has she ever considered getting fast food and using it for content without feeding it to her kids? She's kind of done something like that, she tells me. There was a video where she made snack plates for her kids entirely made up of different types of marshmallows. The video was viewed 2.2 million times, but Alexandra tells me her kids didn't eat the marshmallows. "I knew they wouldn't eat it," she says. "Which I mean—it wasn't just [for] the content. I mean, eventually I did eat [the marshmallows]. So I guess technically speaking, I have rage-baited a tiny bit." Who can blame her? If people are going to hate on her—which they do—and if that hate makes her more money, why not stoke the rage a little? If rage could buy me a house, I would bait people too.

The amount of money in the mom influencer and family vlogging world is almost unbelievable. Whatever you're thinking, it's more than that. According to Goldman Sachs Research, the creator economy is expected to reach $480 billion by 2027, and the highest echelons of mom influencers and family vloggers make hundreds of thousands or even millions of dollars a year.

For example: A few years ago, I ran across @lottie..weaver's videos on TikTok. Lottie is a thin, blond mom in her early thirties who has three really cute young daughters. In 2023, she started posting daily "get-ready-with-me" videos on TikTok where she styled her daughters' hair for the day. In one of her most viral videos, which has been viewed over 32 million times, Lottie wraps her middle daughter's hair around a curling iron as she says in the voice-over, "Yes, we do this every single morning." It was only a few months after she went viral for the first time that Lottie signed with a talent agency. They asked her what her income goal was. "I said $3,000 [a month] would be a dream," she tells me over a phone call. Lottie signed with her agency in June. By August, her monthly income was around $15,000. "I was like, you have to be shitting me. My husband and I both, we were, like, jaws open, jaws to the floor." Now, with 2.2 million followers on TikTok, Lottie estimates that she makes "between $25,000 and $40,000 a month" through a combination of the TikTok Creator Fund, affiliate links, and brand sponsorships. Before she started doing TikTok, she was a cheerleading coach; an entire year's salary in that job was about $25,000. "Never in my life did I think . . ." she trails off. "I just wake up grateful every day and I do my job."

In her best year as a family vlogger, Rossanna Burgos and her family made $1 million. It all started ten years ago, when Rossanna's husband posted Vine videos with their two young children that went viral. Rossanna (known to fans as Mama Bee) remembers asking her husband why so many people were watching their family. "He said, 'People love us. They see themselves in us.'" Once their content started taking off on Vine, Rossanna's husband Andrés (known online as Papa Bee) made the family YouTube and Instagram accounts. They now have 10.4 million YouTube subscribers and 1.9 million Instagram followers.

The Bee children, Gabriela and Roberto, were around nine and ten years old when the family started creating content. The kids loved making videos with their parents, Mama Bee says. Their most popular video of all time has 106 million views and features Papa Bee placing covered plates in front of the kids (affectionately referred to as "the monkeys") who play rock-paper-scissors to determine who is going to eat a gummy food version of the real food the other one is going to eat. Roberto wins the game and chooses a covered plate that turns out to hold a gummy hamburger. Both kids cheer and bite into their burgers. The video continues with several more rounds of rock-paper-scissors and gummy foods including peppers, soda, and pizza. Within a couple of years, the Bee family was approached for brand deals. Their first brand deal, Mama Bee remembers, paid them $500, which seemed "unbelievable" at the time. Outside of YouTube, Mama Bee worked in pharmaceuticals, and Papa Bee worked in market research, both netting six-figure salaries. As their platform continued to grow, there came a moment of reckoning. "And so my husband and I had a very serious conversation of, what do we do? Do we go all in and quit our jobs? Or we don't want to look back when we're ninety years old and say, I wonder what would have happened if we had [gone] all in." So they quit their jobs and dedicated themselves to YouTube. It took a few years before their new social media career exceeded their previous salaries, but in their best year, they made "in the vicinity of one million [dollars]."

Mama Bee says, "It's amazing, but it can be very dangerous in the wrong hands. . . .What I mean by that is that I have seen a lot of families sacrificing the well-being of their family, of their marriage, of their mindset, of their integrity, of their morals in order to try to chase that and achieve that," she says. Her heart breaks for kids who were born with cameras in their faces or for kids shown in vulnerable moments like when they're injured (which both Sam and Julie Jeppson told me is the content that performs best on their channels). Mama Bee sees herself as distinct from these kinds of creators (in my research, I've found every creator sees themselves as different from those at the center of the family vlogging backlash). A quick scroll through the Bee Family's content shows videos mostly focused on challenges and skits.

There aren't videos about either of the children going through puberty or struggles they had in school. And, Mama Bee tells me, she knows the truth behind some of the family vlogging channels who deserve the backlash.

"You go to VidCon, and you will see a lot of these families on stages, and the kids' microphones, and everyone's talking, and it's just this picture of perfection. But then you go backstage, and you've got these, forgive me, narcissistic parents who want the limelight who will do anything for relevance. The kids are like, 'Mom, Dad, I'm hungry. Mom, Dad, I'm tired' [and the parents are like], 'Yeah, yeah, yeah.'"

Mama Bee doesn't want to be seen as a family vlogger, because she thinks of family vloggers as people who vlog every single moment "from morning to night, hoping to capture something that will stand out" as opposed to the more challenge-focused and scripted content on her channel. Meeting other vloggers at conventions like VidCon really solidified her view. "We've also met a family where their daughter pulls her hair out of her head because the parents have a camera in her face at all times, hoping they'll capture something witty and funny and clever. And this poor girl has no idea who she is because she is always on," she says. I ask her how she heard about the girl pulling her hair out of her head, and Mama Bee says the girl's mother told her in a very "matter-of-fact" way. "You're focusing on dollar figures, you're focusing on likes, you're focusing on comments, you're focusing on views. Oh, my views are going down. I need to have another baby. Let's bring another baby into the story, into the picture. That's a human being. That's a child that you need to raise, that you need to go to a university. You have to buy clothes, you have to buy shoes and food. You have to give emotional support. It's not just another character in the game."

In our conversation, I get the sense that YouTube and content creation have left the Bee Family behind. "It's not what it used to be,"

Rossanna says. "Thanks to MrBeast, where he made videos of such tremendous—I'll use the word *sensationalism*—I bought an island and I bought an airplane, and I did this.' You can't compete with that. And that's when we decided to take a pause."

In February of 2024, the Bee Family posted a video titled "Why Our Family Changed . . . ," which currently has only 129,000 views. "Sensationalism killed the creator," Rossanna says while sitting next to Papa Bee. "Now if you're not flying a car off a cliff or eating fifty hot dogs in thirty seconds—" Papa Bee jumps in: "Fifty hot dogs? That's so 2013. Now they're eating, like, four thousand hot dogs in ten seconds." The couple goes on to explain that they're stepping back from constant content creation, but viewers should keep up with their kids, Gabriela and Roberto, who are growing their music careers. (A quick check shows that Gabriela has 10.9 million TikTok followers and Roberto has 246,000.) The family also creates Roblox games, which gives them another stream of income.

Mama Bee tells me that each of their children have money saved in a trust. How much? I ask. She's evasive, calling it a "good chunk of change. What a way to start their lives." "I hate to push," I say, "but I have to ask. Is it hundreds of thousands? Is it millions? Is it tens of thousands? What are we talking?"

"I would say millions," Rossanna says.

So yes, she says. If given the chance, she would do it all again.

Clarissa Laskey, a former mom influencer turned influencer marketing manager, told me that she is astounded by the amount of money influencers make, every single day. "In a giant program we just ran, a creator was paid over $150,000 for one TikTok post." *(One hundred and fifty thousand dollars. For one video! Should I become a mom influencer?)* Brandon Stewart, the founder of Brandon Studios and a former producer at AwesomenessTV, tells me he worked with kid influencers who would make $100,000 a month.

How do influencers make their boatloads of money? For the answer to that question, I reached out to Brendan Gahan, the CEO of the influencer marketing agency Creator Authority. Gahan, who has worked in the industry for fifteen years, sends me an email of notes he's written up for me to look over.

First, Gahan's email explains, there's income from the platforms themselves: the TikTok Creator Fund, which pays creators who have more than 10,000 followers for eligible views on videos longer than a minute, and YouTube Adsense, a program through which eligible YouTube creators are given a 45/55 split in ad revenue in favor of the creator. They're paid based on the CPM, which refers to the cost per 1,000 impressions. Basically: A creator gets a cut of what advertisers pay when an ad is placed on their video (though not all videos have ads and not all views are eligible for payment).

Remember the Family Fun Pack, the YouTube family with 10.4 million subscribers? They have over 15 billion lifetime views, and by the time I speak to Gahan, they've netted 89 million views in the previous 30 days. Gahan looks over their data while we're chatting, calling their billions of lifetime views impressive. I ask him to take a shot in the dark on how much money the Family Fun Pack is making per month just on YouTube from AdSense. "I would think $200,000," he says. "A month?" I ask. Yes, he says.

YouTube and TikTok are the platforms that best compensate creators (though Facebook does pay certain creators, and Instagram used to have a similar program). I wonder: Would Gahan rather have 10 million YouTube subscribers or 10 million TikTok followers? "YouTube, hands down," he says. "YouTube is the most fair from a monetization standpoint. It's very transparent around how you're getting paid, why you are getting paid, what you are. It's also the most amount of money being shared with the partner program. Then on top of that, your content is way more evergreen. Whereas TikTok, the shelf life of stuff is so quick. Then also the relationship in the day you have with your audience on YouTube versus TikTok, it's so much stronger on YouTube. The parasocial relationship is impactful. There's definitely this feeling of a deeper connection there that doesn't exist on the other platforms." Anthony Ambriz, a YouTube strategist based in Utah, told me that the vlogging families he worked with would make tens of thousands of dollars off one video. "They would complain if a video didn't make $20,000," he says. "A bad month would be anything under $50,000."

The second way creators make money is through brand deals and sponsored content. Brand deals are the largest source of income for creators, according to the Goldman Sachs Creator Economy Report. "Amongst creators as a whole, this is still the bulk of where they generate their income," Gahan says. "It powers most of the creator economy." I scroll through my Instagram feed and happen upon a sponsored post from @maiaknight (1.6 million followers on Instagram, 7.6 million on TikTok) for Dove skin products (#GetCozyWithDove). When I ask Gahan how much he thinks Knight was paid for the ad, he notes that she's repped by Digital Brand Architects, a firm he considers to only represent premium influencers. His guess for the 30-second Dove Instagram ad is that Knight made around $50,000 though he wouldn't be surprised if the profit reached up to $100,000, an estimate based on his knowledge of similar brand deals and the number of followers Knight has. And it's not only influencers who have millions of followers who are raking in the money on brand deals. A mom influencer who has just over 500,000 Instagram followers (she requested to be anonymous because she doesn't want people to think she's bragging about the amount of money she makes) told me she made over $300,000 through brand deals in the last year. "But the year before that wasn't even close to half of that so every year is different," she adds.

The third way creators make money is through their own products or businesses. "More and more creators are attempting to use their platforms to launch their own companies (à la MrBeast and Feastables)," Gahan says. I think of the teen mom TikTok creator who launched her own boutique after gaining millions of TikTok followers or the family vloggers who sell merchandise with their kids' faces on it. The biggest creators lend their faces to products, like the Ryan's World–branded toys in Target or Walmart.

Fourth, Gahan says, big creators do appearances, speaking engagements, or conferences, for which they are paid (though this seems to be the least common way for creators to make money).

And fifth, the largest creators can even be acquired, invested in, or given equity in larger companies. Shay Carl Butler was one of the first family vloggers with his channel The Shaytards. Butler had millions of subscribers when he cofounded Maker Studios, a production company that was eventually acquired by Disney for $550 million, a deal for which Butler was rumored to have made $20 million.

How much influencers get paid is not only dependent on how many followers they have. Sometimes, people with massive followings will run brand sponsorships and the return on investment for the brand is low. What "return on investment" means in this context is, for example, the number of people who click the sponsored link the influencer puts in their bio or uses their discount code. "[Number of] followers isn't always indicative that they're going to be good on the program," Laskey says. Is there any way to know if someone will be successful at a brand sponsorship ahead of time, if follower count isn't always a solid indicator? "A lot of times there isn't any way. You're at the hands and mercy of an algorithm."

As the algorithm changes, so does the influencer's income, a former *Bachelor* contestant turned mom influencer who I'll call Rose tells me. She was making over a million dollars a year at her peak around 2018.

Now, she makes between $600,000 and $700,000 annually. The influencer tells me she thinks the decrease in income is due to the whole industry shifting toward sites such as Like To Know It (also known as LTK), where influencers gain commission by providing links to every item of clothing they're wearing and the products they are using. "A lot of that is based around constant hauls and linking, which is something I'm not interested in," Rose says. In 2021, she was approached by the management company of a mom influencer with 2 million Instagram followers. On the call, Rose was presented with the income of the other mom influencer—over $500,000 a month. "I was pretty blown away by that," Rose says. "It did make me feel kinda gross towards her content that was all about how her followers were her besties or whatever and then posting a million links a day. It made me feel, as a viewer, that I was a cog in her money machine."

"There's crazy, crazy amounts of money," says Tyler Chou, an attorney for YouTube creators. It's not only the sponsored content, Chou says—it's the affiliate links where mom influencers and family vloggers can really make money. "Being a family influencer is the best because as the mom, you can do everything, right? You can do fashion, you can do home goods, you can do all the kids' foods. All foods, all foods basically you can touch. You can do kitchen appliances, you can do furniture. Of all the creators, I think being a family channel is probably the most profitable."

The money that can be made off affiliate links is nothing to scoff at. When I caught up with Madison Luscombe, the cofounder of The Creator Society, a management company that specializes in affiliate marketing, she told me that her most successful mom influencer made $3.6 million last year *just off affiliate link commissions*. That same influencer has made $1.1 million in the first quarter of 2025. I find myself shocked by these numbers, but Luscombe isn't. "Moms are the true converters in the [affiliate marketing] space," she says. "They're the ones who really know how to sell things. Women are the household deciders when it comes to what they're going to purchase. Moms trust other moms. Anytime we find out a creator is pregnant, we're like, 'Please do affiliates because you're going to sell so much to other moms,'" says Luscombe, the affiliate marketing agent. "There's so much there." I had to ask: You know those influencers who post affiliate links to multiple products every single day? Most of them aren't doing it themselves. Agencies like The Creator Society take over the backend. Luscombe and her colleagues have the passwords of the influencers, she tells me, and they jump on and post the affiliate links.

From the beginning of your transition from an influencer to a mom influencer or a vlogger to a family vlogger, there are so many sponsorship and affiliate opportunities. Trying to get pregnant? Do a Modern Fertility brand deal detailing how you're tracking your cycle to pinpoint ovulation. At the point of taking pregnancy tests? There's a Clearblue sponsorship waiting for you to post a crying video finding out you're pregnant using one of their tests. Packing your hospital bag? Link to everything you're bringing and swim in the affiliate cash. And then when the baby is born? The opportunity for content explodes—you can link to your favorite diaper brand and do sponsorship deals for baby clothes and bottles and formula. Consider @alyssafluellen, a family vlogger with 6.4 million YouTube subscribers, 5.1 million TikTok followers, and 1.8 million Instagram followers, who announced the birth of her fourth baby on Instagram with the caption "She's here and she's perfect. Pjs: @silks.design #laboranddelivery #newborn #announcement."

As the kids grow, so do the opportunities for commissionable content—you can link to sippy cups and front-facing strollers and your favorite toddler outfits. You can set up brand deals for their supposed favorite snacks or Target-sponsored back-to-school shopping trips. And it's not only kid-specific stuff—moms can do deals for any product related to running a household which is, uh, almost every product. I once saw a mom influencer make a sponsored post for batteries. "You can literally do anything," Chou says. "Everything can be touched with a family channel, basically."

Once vloggers and influencers start to make real money, Anthony Ambriz, the YouTube strategist, says they offload the work of their channel—and household. "They hire editors from the Philippines to cover editing and channel management," Ambriz says. "They have editors, videographers, household managers, nannies, housekeepers, managers for brand sponsorships." Ambriz used to hold creator meet-ups where prominent family vloggers would get together to discuss the best strategies and content ideas. At one of these meetings, Ambriz and a vlogger dad were preparing a presentation to give to the rest of the group; Ambriz asked the dad what he thought the title of the presentation should be. The dad's suggestion? "How to get rich exploiting your kids."

Among vlogging families, it's common knowledge what type of content does best, Ambriz says, including content where a child is sick or hurt and content surrounding a pregnancy or the arrival of a new baby. "There was that jealousy, where someone would be like 'I'm not growing as fast, but they are. Their kid keeps getting sick, of course they're going to grow.'"

It's not only sick kid content that does well—generally, content with children tends to do better than content without children. Clarissa Laskey, former mom blogger turned social media marketing manager, says she noticed this herself when she was a mom creator. If she was traveling without her kids, people would always comment asking where they were. And when she posted with them, those posts did better than her solo posts. Research backs Laskey up: Pew Research Center found that YouTube videos that featured a child or children under the age of thirteen were substantially more popular than videos that didn't feature children.

Because of her past as a mom influencer and her present as an influencer marketing expert, Laskey is the perfect person to ask the question I've always wanted to know the answer to: How much money are these people actually making? She estimates that the LaBrant family, through YouTube views and sponsored posts and the TikTok Creator Fund, are making in the range of $3 million annually.

"They're probably demanding anywhere from $200,000 off a [sponsored] post, if not more," she says. "They're also making YouTube ad revenue. So an ad revenue of another $10,000 to $12,000 a month from Google ads on YouTube."

For YouTube creators with 10 million subscribers, Laskey estimates they're making between $5 million and $8 million annually between YouTube ad revenue and sponsored posts. Even people with only 500,000 subscribers can make $6,000 a month on ads alone.

Laskey estimates that Maia Knight, the twenty-eight-year-old mom of twin daughters Violet and Scout, who has 7.7 million followers on TikTok, could be "in that three-million-to-five-million-dollar range a year annually," she said. Knight no longer shows her daughters' faces on any social media platform, which she called "a choice to protect [them]" in a TikTok video explaining the decision. According to Laskey, almost all creators who historically showed their kids online and then decided to take them off made less money after taking their kids offline. How much money were they losing? I ask. "Half their money," she says. "Let's say you're making $100,000 a year, you're probably now down to sixty. Maybe fifty. You can charge a lot more [with kids]." And sometimes the reality can be even more stark than that. A source I interviewed for a story for *The Washington Post*, Grant Khanbalinov, the dad behind @lifeofbreaandgrant on TikTok (3.1 million followers) told me that once he stopped showing his kids in content, he went from making $100,000 a year in sponsored content to "pretty much zero. Since we stopped, we've gotten one brand deal in the last year," Khanbalinov told me. "There's no amount of money that someone could pay me right now to say, 'Hey, put your kids in a video right now for TikTok.'"

Katie Beach, a mom influencer with nearly 150,000 followers on Instagram and 114,000 followers on TikTok, doesn't show her kids online. If her toddlers are in her videos, they're filmed from the back or side. "I hate the idea of strangers being able to recognize my child," Beach says. But that protective instinct has cost her. "I've turned down numerous five-figure brand deals because I wouldn't show my children," she says. "I've had diaper brands, for example—not only would I not show my kids, I would never show my kids in a diaper. There are baby food companies who want to show your kids eating, which is another thing I would never do. And just so many toy companies that want to show your kids playing with the toy." Usually, she says, brands will reach out with an offer; included in the offer is a brief describing their vision of the content. "It'll be like, 'We want you to show your child doing X and Y,'" Beach explains. "Or 'We want this kid to be in diapers.'" Beach responds to the offer, saying she doesn't show her kids online, but she's happy to approach the content in a different way, like showing the back of her toddlers' heads. "And brands pass," she says. "I've had email chains where the brand has said, 'Alright, since you're not going to show your children, we'll be passing because that is a requirement for this creative.'"

Though more mom influencers are choosing to keep their kids offline in recent years, some brands just aren't open to ads or sponsored content that doesn't feature children, says Monica Banks, the founder of Gugu Guru, a marketing firm that connects mom influencers with paid opportunities. And for brands, there's no lack of creators to pull from. "There are still plenty of moms who are okay with showing their kids in the content," Banks says.

Ambriz also mentions that pregnancy and newborn content can be particularly lucrative. "And then it becomes this joke of, should we have a kid so we can start growing [our channel]?" Ambriz says. "I think it's a joke that sometimes stems from reality."

Laskey, the mom blogger turned marketing manager, addresses this oft-repeated rumor—that vloggers and influencers have kids specifically to create content—that has been running rampant through the family vlogging world for years. "I hate to say this, but over the years, I've known people who have had more children because those brand deals are really lucrative." It stops me in my tracks: "Is that true?" I ask. "I definitely know some people," she says. "I've definitely been in circles with certain influencers [where they're] all of a sudden adding a fourth or fifth child. Or before their baby's even here, they already have brand deals lined up and what they can get and how they can get their nursery paid for now." There's so much money in the baby industry," Laskey says. "In the baby world, everything from clothing to all the things now they can sell for kids. Every baby-wearer, every sling, every stroller, that stuff can be shown. You can be a mom traveling and show in a post a stroller, a pair of socks, an outfit, a bottle." The baby industry is huge—and only getting bigger. In the 2024 paper, "The Use of Children as Influencers and the Harmful Effects on Their Health and Rights as Human Beings," researcher and lawyer Ingrida Behri wrote, "In the last two decades, the children's market has expanded dramatically. This in turn has led to increased attention to children by marketers and to a process of commodification of childhood." According to a Market.us report, the global baby products market has grown to $361.2 billion in 2025 and is expected to reach $575.8 billion by 2033.

I can't help but keep returning to this: Does Laskey really think that people are explicitly choosing to have more children for brand deals? Imagine choosing to bring another entire *person* into the world for brand deals. It boggles the mind. It's almost unbelievable—except I've heard it from several different sources who have spent their professional careers in the influencing industry. I ask her how explicit the thinking of the influencers is. Is it literally *I need to have another kid because it is good for business*? "That is definitely a thing," Laskey says. "It's a thing of saying, 'You know what? This was a great ride. Now my kid is six years old and no one cares as much. You know, maybe we should have another baby and get these brand deals going.'" She continues, "I personally don't think it's great because those little kids that you have with the great brand deal, they grow up . . . and those deals go away and now you're still raising a child, so there's still a cost and your time and energy and all the things of being a parent. And you have no idea—I've always said this industry is very fickle. It can be very lovely and when you're riding the wave, the wave can be great. That wave can crash very suddenly, and everything you thought you built can be taken away in a second." But while it's good, it can be great. "There are family vloggers that are so large that they live in gated communities next to NBA players and CEOs of airlines," Ambriz, the YouTube strategist who worked with prominent family vloggers, says. Dani Abraham, who used to be the senior manager of talent partnerships at AwesomenessTV, a media company that cultivated kids' YouTube channels, says, "These kids would rake in $100,000 a month."

When we talk about the family vlogging industry and what it's like to be an influencer kid, we need to talk about the money. This is life-changing money. I know there's a bleak lens through which to view this: Parents are exploiting the privacy of their children in exchange for cash. In some cases, that might be true but that's not the whole story, and ignoring the complexities of this industry doesn't do the kids at the center of it any justice. I think of Bethanie Johnson of @thegarciadiaries, with 322,000 followers on Instagram, who makes about $500,000 a year. Though she's considered pulling back from influencing, she knows there's no other job in which a single mother of five with only a high school education could make half a million dollars a year.

Or MariClare, a teen mom who has 2.2 million followers on TikTok, who makes between $10,000 and $30,000 a month, and tells me she's never had a job outside of content creation besides briefly waiting tables.

One family vlogger told me they netted about $360,000 in their best year; before family vlogging, they were living paycheck to paycheck.

Kaci, who with her husband, Casey (yes, both their names are Casey, and they're high school sweethearts) has 1.6 million followers on TikTok, didn't reveal exactly how much money the family makes but offered the information that her husband was an exterminator before they took off on social media. When their account first went viral, they were making "three times that. Now it's a whole different world to us." I ask her what that freedom feels like. "Honestly, it's nice not to have to worry about bills or food. There was a time when we had to choose between getting gas in our car or paying a bill. It's so nice not to have to deal with that anymore."

Adrea Garza, the mom behind @garzacrew with 5.2 million TikTok followers, features the lives of her twin eight-year-old daughters Haven and Koti. When Adrea found out she was pregnant with twins, she realized that if she continued to work, her entire paycheck would go to daycare costs. Instead, she decided to stay home (although how much of a choice did she have?). Now, five years after her daughters first went viral, the girls' social media presence is Adrea's full-time job. "I'm constantly on the phone, Zoom calls, talking to friends to brainstorm ideas for content, talking with our accountants, our managers, my attorney. My day is pretty much spent managing this Garza Crew media company that I've now created." Adrea estimates that this year, she and her daughters will rake in somewhere between "$500,000 and a million, just depending on how the year shakes out."

Though influencing is a global career, it seems to me a uniquely American pursuit. It appeals to the best parts of the American Dream, the ones we can't help wanting to believe in: that anyone can be a success and have the white picket fence house and three-car garage to show for it. In a country where economic stability feels so precarious and increasingly unreachable for all but the wealthiest among us, is it any wonder Americans are taking their turn playing the viral lottery? In late-stage capitalism, where everything is a commodity, including you and your child, there's a part of me that understands trading privacy for economic stability.

In an ideal world, childhood would be sacred, existing outside of the bounds of being turned into profit. But we're not in that world—we're in this one, where there are fewer and fewer paths to the ever-shrinking middle class, where childcare costs often outpace a parent's entire salary, where student and medical debt continues to balloon—and where one sponsorship can net more than the average yearly salary. I wonder if, when people rage against influencer parents, what they're really raging against is the system that offers so few opportunities for advancement besides the commodification of the self. All you have to do is post.

12

The Wrong Kind of Attention:

Online Predators and Invisible Threats

In 2022, Wren Eleanor, who was three years old at the time, became the focus of a social media firestorm as thousands of people called for her mother, Jacquelyn, to remove content featuring Wren from TikTok, where the duo had nearly 17 million followers. TikTokers latched on to Wren as the face of the issue of child exploitation. Why, they wondered, was her mom posting videos of her that could be seen as suggestive, like Wren eating a hot dog or saying "I swallowed it" or pretending to breastfeed her stuffed animal? "It was like bait for creeps," said Taylor Lorenz, an internet culture journalist and author of the bestselling book *Extremely Online*. There was even a video where Wren pretended to shave her pubic region and insert a tampon as part of a trend where parents gave their kids everyday items to see if the child knew how to use them, according to Sarah Adams, the creator behind @mom.uncharted, a popular account on TikTok dedicated to sharing information on what Adams sees as the exploitation of influencer children. "It just felt so unbelievably inappropriate that a mother would film this, put this all online," Sarah said. "I think that was the video that really caught the internet's attention."

That's where the online sleuths came in. On a subreddit dedicated to Wren and Jacquelyn, anonymous Redditors claimed videos of Wren in a bathing suit or playing with a tampon got way more saves and comments than other videos that could be seen as less inappropriate. According to *Rolling Stone*, Redditors "even tracked down the social media footprints of some of Wren's followers, pointing to comments they have made or other videos they have saved to indicate they may be sexually attracted to children."

After she saw the video of Wren pretending to shave her pubic region, Sarah Adams took matters into her own hands and contacted companies that worked with Jacquelyn. "I reached out to a company that she was affiliated with at the time and I said, 'How are you supporting this? This is crazy,'" she said. The company responded that they weren't moving forward with further partnerships with Jacquelyn.

Something about Wren, a toddler with 17 million followers who seemed to have been filmed almost constantly for years, struck a nerve with parents, many of whom took to TikTok to urge others to delete photos of their children, lest they be saved by pedophiles for nefarious uses. The discussion around Wren's content broadened to include not only the possible pedophiles lurking on the internet but also the rights of child influencers generally. Jacquelyn didn't comment on the controversy, but sometime in 2023, she took her account off TikTok and hasn't been heard from since. When I reached out to TikTok for comment on the scandal, a representative pointed me toward their community guidelines, which bar content that puts "young people at risk of psychological, physical, or developmental harm. We do not allow content by young people that intends to be sexually suggestive." Similarly, Meta's community guidelines include protections against content that "sexually exploits or endangers children" regardless of who is posting it. But what I think these platforms haven't grappled with is what to do when possibly suggestive content is being posted by a parent. Though they do explicitly say they do not tolerate child sexual exploitation and abuse of young people under the age of eighteen, it seems like the content that Jacquelyn posted of Wren straddles a line that can be difficult to decipher. Is it child exploitation and abuse to post your three-year-old daughter pretending to insert a tampon? Jacquelyn's page is currently blank and doesn't show any of her old posts. I wondered if TikTok had taken her content down, but when I asked, they said they hadn't taken any action on her account. They did tell me that search terms including the phrase "Wren Eleanor" have been banned by the app.

When I spoke to journalist Taylor Lorenz about the Wren Eleanor controversy, I asked her if she thought the fever pitch the controversy reached was overblown or reasonable.

"Well, I think it's a little bit of both," she said. "I think that the videos were weird enough to get attention. We need to have those discussions and be able to call out weird content. That said, I think some of it is a little bit overblown in the sense of, is the problem children on the internet or is the problem creeps? Should parents post their kids less on the internet? Sure, probably. Should a mom be uploading these weird, suggestive videos of their toddler? We can agree that's stupid and bad." But, Lorenz argues, the answer requires a societal reckoning about "our pedophilic culture and why so many men consume this content or are interested in it. To me, that's the bigger problem and that never gets discussed in these outrage cycles because it's always focused on the mother and demonizing her. It provides everyone this easy outlet to hate on a woman for a while, but it doesn't do anything to fix the broader systemic problem."

And what does Lorenz see as the broader systemic problem? "The misogyny of pedophilic culture that fetishizes children and especially young girls."

In February of 2024, *The New York Times* published a story called "A Marketplace of Girl Influencers Managed by Moms and Stalked by Men." For the investigation, the *Times* reported that they analyzed 2.1 million Instagram posts, interviewed over one hundred people, including parents and children, and "monitored months of online chats of professed pedophiles."

When I read the story, I was shocked. It included harrowing details like pedophiles cheering on the publishing of kid influencer content. ("I'm so glad for these new moms pimping their daughters out," wrote one in a chat room dedicated to the topic. "And there's an infinite supply of it—literally just refresh your Instagram Explore page there's fresh preteens.") Time after time, the social media companies in question (mainly Meta) allegedly failed to act when moms reported pedophilic comments or messages. This excerpt sticks in my head still, a year later:

> If parents block too many followers' accounts in a day, Meta curtails their ability to block or follow others, they said. "I remember being told, like, I've reached my limit," said a mother of two dancers in Arizona who declined to be named. "Like what? I reached my limit of pedophiles for today. OK, great."

I reached out to Meta for comment, and they told me the company not only bars explicit sexualization of children but implicit sexualization. "We remove accounts that are dedicated to sharing otherwise-benign images of minors, when the captions and comments are predominantly focused on the children's appearance," the representative said, noting that Meta reports all apparent instances of child sexual exploitation content to the National Center for Missing & Exploited Children.

The *New York Times* story confirmed peoples' worst fears: Pedophiles were a genuine threat to their children online. And pedophiles considered the content parents posted online to be a virtual treasure trove. Still, despite this evidence, some mothers didn't choose to change the content they posted. One mother in the story described herself as being desensitized to the men's messages. "I don't have much of an emotional response anymore," she said. "It's weird to be numb to that, but the quantity is just astounding."

Detractors of vlogging families and influencer kids often claim that there is a genuine risk of child influencer content being utilized by pedophiles or turned into child sexual abuse material (CSAM). In a conversation with Kevin Smith, a public affairs officer at the Federal Bureau of Investigation, I ask directly: Is that a risk? "It is a risk for sure," he says. More than a risk, it's a reality: A 2025 study that interviewed nineteen influencer parents found that "every interviewed parent, except one, had dealt with pedophiles. Particularly, several mothers have received messages, photos, and/or videos from pedophiles, sometimes on a very regular basis." So why do they keep posting? "[Parents] are incentivized financially to do it, and that might supersede a safety matter," says Smith. "Some of the stuff that's out there, it never goes away. And it could come back and be a predator's dream. Half of the FBI might go out of business if the internet went away. We can hardly even keep up with the number of platforms and ways that these people can get into children's lives." Additionally, researchers have found that "innocent content published on social media may appear in other contexts for which it was not intended, such as pedophile networks, and make children a potential target of child predators." And the problem is only worsening: In July 2025, *The New York Times* published a story that stated that "a new flood of child sexual abuse material created by artificial intelligence is hitting a tipping point of realism, threatening to overwhelm the authorities."

In the course of this reporting, I spoke to an individual I'm going to call Carrie who works with people in post-release reentry programs following serious offenses. (Carrie requested to remain anonymous because she wasn't cleared by her bosses to speak to me.) Sometimes, Carrie told me, when offenders were released, they weren't given digital repercussions because their crimes weren't digital in nature. To make this simpler: Let's say a person is convicted of assault—their post-parole guidelines may not include limits on phone or internet usage because their assault happened in person rather than online. Part of Carrie's job is checking up on people going through reentry programs. She told me of several instances in which she or another investigator went through the phones of offenders and found saved images and videos of children from YouTube, TikTok, Instagram, and other platforms. "All of it was public information. They had all this access to children," Carrie says. "By the time one man left our facility, he had over a thousand saved photos. He found these kids through Instagram, through YouTube, through public sites that influencers are posting their kids on." For Carrie, the calculus is clear: Her child's face and name have never been posted on the internet. "You really have to weigh your risk versus reward," she says. "What does that look like for you?"

Adrea Garza, the mom of eight-year-old influencer superstars Haven and Koti, who have 5.2 million followers on TikTok, says the prospect of predators watching her content is something she tries not to fixate on. "My daughter does dance. Could there be a pedophile in the audience at her dance competition, watching and filming these little girls? Absolutely. I think it's a shame, and I think it's awful that these people exist in the world, but I can't put my kids in a bubble."

Plus, she adds, "What I think you really need to be more conscious of is that's probably going to happen with an uncle or a cousin before it's going to happen with some stranger online. I do my best to just always keep them safe and protected, and there's only so much that we can control in this world. I do what I can do while still allowing my kids to experience life and their passions." Garza once told me that, because she feared creeps were watching the content she posted of her daughter dancing, she limited how often she posted that content. "I might post it once a quarter," she said. "Because 95 percent of our followers are little girls that just are fans. And so if there's some sicko out there that twists and is weird about stuff, he's going to do that whether we're online or we're at the park and he takes a picture."

The logic here is flawed to the point of senselessness. Because a pedophile could be lurking in a public park, you might as well post your kid online in flimsy clothes? And if you're worried that photos of your young daughter in her dance costumes get unwanted attention, how does it solve that problem to only post those photos a few times a year?

I don't know what to make of the defenses of the parents I asked about this because, according to Sam (the family vlogger parent who bribed their kids to participate in content), they know pedophilic men are lurking on their accounts—and they're coming right up against the line of catering to them. "There's no secret in marketing and media in general that sex sells," Sam tells me. "Sex always sells, it attracts audiences. And so what ends up happening is when parents try to skirt that line between exploiting and benefiting off that well-known fact, but without crossing the line into what would be largely deemed as inappropriate." The AI-powered algorithms that control the virality of content have no sense of morality, Sam adds. If content is being watched, AI doesn't care if it's being watched because it involves the sexualizing of children. It only cares that it's being watched—and it will push it to more viewers.

There was one family in particular that Sam was close to during their time in the family vlogging world. The family, whom we'll call the Smiths to protect their identity, had a daughter we'll call Callie, who was around eleven or twelve years old. Sam remembers watching as the

Smiths started posting pictures and videos of Callie, in "really short miniskirts and tight-fitting tops, tank tops. And she was all decked out in makeup, and she was standing there doing these kind of like sultry, open-mouth poses." The posts featuring Callie were doing really well. The comments were full of sentiments like "She's so beautiful" and "I wish I could look like that" and "What a gorgeous girl." Though the Smith family got a surge in followers seemingly related to the Callie content, there were some comments that rubbed them the wrong way. Some commenters accused the Smiths of posting pedophilic content, while other comments seemed overly sexual. "And I remember [the parents] were just furious," Sam says. "I mean livid at those types of comments. They would delete them as soon as they could. One because they wanted to keep it under control. Creators look at comments as weeds, and if you don't keep the comments weeded, then they result in more weeds."

Were the Smiths deleting comments because they thought there was some truth to them? Or because they were so offended by the accusation? Because they knew there was some truth to them, Sam says. "Every parent of a [kid] influencer knows that creepos are out there, and they don't care. They justify it and say, no, the platform's safe and you know, yeah, creepos are out there, what can you do? They say all that kind of stuff. We said all that kind of stuff."

Once the Smiths started pushing the boundaries of the kind of content they would post of their young daughter, their growth was explosive. "They became very, very popular," Sam says. "But the problem is what it's like feeding the beast once you start posting that kind of content. In order to keep the views coming, the clicks coming, the subscriber and follower growth coming, you have to keep uploading more and more of that type of content." And that's what the Smiths did, according to Sam's telling. Then, one day, Sam and their spouse were hanging out with the Smith parents, who were furious and venting because the website OnlyFans had reached out to them with a personalized message telling them that when Callie turned eighteen, if she joined the website, she would make several million dollars a year.

"And they were so livid," Sam remembers. "They were like, 'Where would they get that idea? Why would they reach out to us? Has OnlyFans reached out to any of you before? Like, why are they singling us out?' And the rest of us were kind of, like, rolling our eyes, you know, because we knew. We knew exactly why."

In the Kids as Content Digital Safeguarding Toolkit project at the University of Essex, Dr. Francis Rees, a lecturer in law, writes that the sexualization of children's content is one of the potential harms facing influencer children. "Research evidences that images where children are wearing swimwear or underwear are saved and shared more often than any other type of child-related content. Additionally, apps like FaceSwap and nudify can be used to manipulate these images into child sexual abuse material and be shared and exchanged with other users," Rees writes. "Similarly, other sources report that for some female child influencer accounts, 73 percent of their followers are adult males and that, where adult users have shown an interest in sexualized images of children, the platform algorithms will actively push child influencer content into their feed. This can also be found and magnified through searching for specific hashtags like #babygirl and #daddysgirl."

It's not that influencer parents aren't thinking about this. When I spoke to Lottie Weaver, who has 2.7 million TikTok followers and three young daughters who are featured prominently in her content, she tells me the prospect of predators watching her content is "something that stresses me out one hundred percent." There have been times, Lottie says, when the worry has made her question whether she wants to keep posting her kids online. But then she thinks, even if you're just posting to your seven hundred friends on Instagram, someone could have the wrong intentions. "If someone is going to sexualize a child, someone's going to do it, whether it's your brother, whether it's your grandma, whether it's a stranger. You know what I'm saying?" she asks. She's quiet for a moment before she speaks again. "Maybe that's a way for me to justify it in my mind." In the study "Work It Baby! A Survey Study to Investigate the Role of Underaged Children and Privacy Management Strategies Within Parent Influencer Content," researchers found that "parents may minimize the risks related to influencer sharenting behaviors to resolve conflicting beliefs and behaviors. By minimizing privacy concerns, they can rationalize their behavior and reduce the tension that may result from the discrepancy between belief and behaviors." Another study found that momfluencers perceived privacy risks as "relatively abstract and distant because the majority of them have not (yet) personally experienced them."

For some influencers, there's more than the suspicion that people may be looking at your content with ill intent. Once, Bekah Martinez (@bekah with 860,000 Instagram followers) shared a photo of her nursing her baby and toddler at the same time. She wasn't covering up, because she thought it was important to show the reality of tandem nursing. "It's important to represent this aspect of motherhood," she says. "I wasn't thinking about pervy people." But after posting the photo, Bekah's followers sent her screenshots of the photo reposted on "some weird Facebook account. It was getting millions and millions of views. I just felt so sick to my stomach. And it was a post that I had on my Instagram. The idea that somebody could be jerking off to a picture of me breastfeeding my kids . . . that's super disgusting, and I regret posting that for that reason."

Jess Spaulding (@jess.spaulding with 156,000 TikTok followers) deals with the threat of creeps on the internet by making sure that when she posts videos of her toddler daughter in just a diaper, there's an emoji covering her chest. "That's something I'm really big on," she says. "I always make sure that she's covered up." (When I asked someone from the FBI, whose name I'm withholding because they were not authorized to speak on this subject, about the practice of covering body parts with emojis, they said: "That's just stupid." They added that the act of covering a body part with an emoji was equivalent to admitting that it would be sexualized. "I wouldn't post it at all," they said.)

Lottie, the mom we discussed earlier who has three daughters and over 2.7 million TikTok followers, doesn't post photos of her kids in the bath. If her daughters are eating popsicles or lollipops, she makes sure they finish eating before she takes a video or photo of them. She also doesn't post her daughters in swimsuits, she says, except for "sometimes on Instagram, if it's a photo or whatever. But videos normally, I don't do them in swimsuits."

Parents know that the content they're posting of their children is being watched and consumed by people with pedophilic intent, according to Elisabeth Van den Abeele's research. In Van den Abeele's thesis for a PhD in communication science, she interviewed mothers of influencers, including one whose child's content, address, and school information had surfaced on the dark web. The child was even stalked at school, filmed by a man who was a fan of hers online. "What I saw was that these mothers were actually very aware of the risks," Van den Abeele said. "That was something that bothered me because especially the girl that went on the dark web, what could happen more for you to decide to just stop doing it? It was really something that triggered me. Why would you continue to depict your children online when you know these risks and when you already experienced certain consequences of these risks?"

In further research, Van den Abeele examined the unconscious cognitive biases that parents (specifically mothers, who are often at the helm of child influencer accounts) employ to justify their decision to continue creating content. For instance, Van den Abeele says, the mothers would say "There are thousands of children on social media—why would a pedophile choose my child to do something with?" Or there was the justification that everyone posts their kids online, so they weren't doing anything different or wrong. A third justification was the illusion of authority: Mothers told Van den Abeele that they deleted followers and comments on a daily basis, which allowed them to feel in control of who was consuming their content. "These things actually lead to the process of them believing that they can actually protect their children from certain negative experiences and actually motivate them to continue [posting content]," she said.

What happened to the mother of the child whose content was found on the dark web? Van den Abeele tells me the mother's first reaction was to take the child offline, which she did. "She deleted the account. She didn't want anything to do with it anymore. But then she argued to me, 'My daughter got really sad and she really wanted to continue being online.' Then two weeks later, she decided to restart the profile and be on social media again."

The evolution was difficult for Van den Abeele to wrap her mind around: Was this mom and others like her just bad mothers who cared more about money than their kids? Van den Abeele doesn't think so—the mothers she spoke to seemed like genuine caregivers who loved their children very much but had just been swallowed into the storm of social media. They started posting, then they got followers, then they started making more money than they could've dreamt of. With their new socioeconomic status, the calculation of privacy and risks versus benefits was hopelessly tilted, so they kept posting.

I don't think we'll ever know for sure how many people are consuming kid influencer content with pedophilic intent. But a quick scroll through the comments of a prominent child influencer makes my stomach turn: “This girl looks 20, not 10,” reads one. “I hope she knows how beautiful she is,” reads another. A third, from someone with a username and profile picture that leads me to guess they’re a man in their fifties: “Very beautiful young lady wish you all the best for the future.”

13

Lights, Camera, Legislation:

Regulating Family Vlogging

Imagine this: You first went viral as a toddler when your parents posted a video of you and your siblings on social media. With that first brush with virality, your parents saw an opportunity. They focused on content creation, roping you and your siblings into scripted videos and dramatically filmed personal moments. They keep posting and you keep working. Your childhood passes in a blur of YouTube stunts and social media fame. You and your siblings work through adolescence, and your family brings in hundreds of thousands of dollars a year through brand deals, sponsored content, and social media payouts. When you reach the age of eighteen, there is not a single dollar set aside for you. In forty-five states, this is completely legal.

In 2023, Illinois made history as the first state to pass a law protecting the rights and profits of child influencers. According to the Associated Press, the Illinois law will "entitle influencers under the age of 16 to a percentage of earnings based on how often they appear on video blogs or online content." That money must be held in a trust that the child can access when they turn eighteen. Illinois Governor J. B. Pritzker's spokesperson Alex Gough told me, "The internet provides more opportunities for children to display their creativity than ever before. In the event that minors are able to profit from that creativity, they deserve to be shielded from parents who would attempt to take advantage of their child's talents and use them for their own financial gain." The law was considered groundbreaking. After months of discussion and pressure from advocates, the profits of child influencers were finally safeguarded—at least in one state. Though the progress was welcome, I couldn't help but think how late it seemed to come, with the Illinois law being passed nearly two decades after family vloggers and mom influencers started making monetized content featuring their children. And the law only applied to kids in Illinois, leaving children in the other forty-nine states without protections until May of 2024 when Minnesota followed Illinois's lead.

Similar to the Illinois law, the Minnesota legislation dictated that if a child aged fourteen and older is featured in more than 30 percent of the compensated content within a 30-day period, then a share of the profits must be set aside in a trust account for that child. If a child under the age of fourteen is featured in monetized content, they must receive 100 percent of the profits. Additionally, if a minor over the age of thirteen requests that content featuring their likeness is deleted, that request must be granted, according to the legislation.

I talked to Senator Erin Maye Quade (D–MN) about what inspired her to pass the legislation and she told me that my reporting for *Teen Vogue*, among other things, served as the catalyst. After reading stories featuring anecdotes from influencer kids and becoming a mom herself, she says she wanted to extend protections to influencer kids. "Adults in power have reacted so slowly to the changing of the landscape when it comes to tech," Quade said. "So the lack of protections didn't surprise me." In Minnesota, legislators were able to extend the existing legal framework for child labor to encompass child influencers. Quade explains that state law dictates that children under the age of fourteen can't work, so the new legislation categorizes content creation as work.

"If you've seen some of the schedules that these kids have to live by or how much content is being put out about them or with them, it is working." The law is triggered by a threshold of a minor appearing in 30 percent of monetized content when at least $.01 is earned per view. That way, if a parent posts a one-off viral video, the law wouldn't apply, but if a person were continually posting monetized content featuring minor children, the law would be in effect. Ideally, Senator Quade tells me, the federal government would pass legislation addressing the rights and profits of influencer children.

When I talk to Senator Quade, I also have to ask: Does she share her own daughter online? "I do not," she said. People can be weird to female politicians online. When she first announced her campaign at the age of twenty-eight, an anonymous man sent her a dick pic. Another man took a photo of her while she was on vacation and posted it online. "So if people are weird with me like that, they would be weirder with my daughter." These anecdotes clarify something for me about my own desire to keep my daughter offline. Once, a few years ago, I got a strange piece of physical hate mail that I never quite stopped feeling weird about. I've been pretty lucky when it comes to backlash online (and through snail mail), but I have gotten some unnerving comments, emails, and messages. Why would I want to give harassers any more ammo, let alone access to the person who is most precious to me in the entire world?

Currently, there is no data that determines which state is populated by the most child influencers, but experts point to California and Utah as two of the hotspots. In September of 2024, California Governor Gavin Newsom signed into law Senate Bill 764, making California the third state in the country to protect the profits of child influencers. Governor Newsom was joined at the signing by pop star Demi Lovato, who just weeks before had released a documentary called *Child Star* that explored the pressures of being a child actor. Similar to the other laws passed, SB 764 dictates that content creators who feature minor children in at least 30 percent of their content must set aside a proportionate percentage of the earnings from that content to be held in a trust account until the child is eighteen. I spoke to California State Senator Steve Padilla, who sponsored the bill, and he told me the legislation was modeled after California's landmark Coogan Law, which in 1939 became the nation's first law to protect the profits of child actors. "Protecting children from all forms of exploitation and abuse is one of our highest callings in government, but our laws have not always kept up with emerging technology," Senator Padilla said. "In the early twentieth century, California took action to protect child actors from financial abuse. With SB 764 now made into law, we expand California's landmark labor protections to twenty-first-century performers as well."

After California passed SB 764, advocates looked to Utah as a state in need of protection for child influencers, especially in the wake of the Ruby Franke case, which served as a spotlight for the pressures facing the children of influencers. In October of 2024, Utah Senator Doug Owens opened a discussion during an interim session of the Utah legislature to gauge support for a potential bill addressing the protections of influencer kids. I broke the news for *The Washington Post*. Though the bill was still being drafted when Senator Owens presented the topic at the session, Owens told me that he had been discussing the bill's language with Ruby Franke's estranged husband, Kevin, and her eldest daughter, Shari. In front of the Utah State Legislature and in one of her first public statements since her mother's arrest on child abuse charges, Shari spoke of the dangers of family vlogging:

"I don't come today as the daughter of a felon, nor a victim of an abnormally abusive mother," Shari began. "I come today as a victim of family vlogging. At first, family vlogging is an alluring business that can bring high revenue. For my family, it became the primary source of income, and this is often the case for full-time family vloggers. Many child influencers are paid for their work. I certainly was, and the money has helped me in my adult life. However, this payment was usually a bribe. For example, we'd be rewarded one hundred dollars or a shopping trip if we filmed a particularly embarrassing moment, or an exciting event."

In her statement, which Shari posted in full on Instagram after the hearing, she continued, "I did not realize the impact that filming as a child would have on me now. My social media became flooded with rumors of having sexual relations with my own brother, to being called a baby birthing machine at the age of thirteen. All these things have stuck with me, and I will forever live from the ages of thirteen to seventeen in many of our viewers' minds. In addition, pedophiles stalk the internet, specifically seeking out child influencers. I promise you that the parents are aware of these predators, and they choose to post their children anyway. If I could go back and do it all again, I'd rather have an empty bank account now and not have my childhood plastered all over the internet. No amount of money I received has made what I've experienced worth it."

Kevin Franke, Shari's father and Ruby's estranged husband, also advocated for the Utah legislation following his wife's arrest. When I talked to Kevin on the phone, he spoke in a measured, focused way, like he had been thinking about these words for a long time. He sounded like the engineering professor he was before his family became tabloid fodder. Kevin knew the tragedy of his family was one of the reasons Utah politicians were even thinking about passing legislation related to child influencers, and though the entire situation is painful to consider, he wonders if this legislation protecting other kids from the fate that befell his own could possibly be a silver lining. "Nobody wants the tragedies of their lives to be used as the justification for making social change, especially tragedies that were as horrific and nightmarish as these were. But I will say this, that for me, personally, it feels a little bit like a work of redemption, in a way," Kevin shared.

Kevin told me he knows that as a parent who put his children on YouTube and chased social media fame, he is part of the problem. He gets that. But he's insistent that he—and the other family vlogging parents he warns me about—are not the only people who should shoulder the blame. "While the spotlight should be shined on [parents], they are only one element of a child exploitation machine that is currently worth hundreds of billions of dollars today. The other element that people aren't really talking about is the role of the social media platforms themselves in incentivizing and promoting content that exploits children." The focus of legislation and activism, Kevin says, needs to be broadened to include the social media companies themselves. "There is no secret among creators and platforms and corporations and those who do analytics in this space, that the secret sauce to increasing views and increasing interactions on these types of videos is adding children to the content."

Kevin hopes future legislation will extend protections past monetary considerations. "In the influencer circles that I hung around in, financial compensation to the children was not the problem," he said. "It was the least of the problems. In fact, extreme financial compensation was often the carrot that was used to incentivize and manipulate the child to interact with the camera or to share things about their lives or to do degrading and embarrassing challenges on camera. And I would ask the question, Well, how much does a child's dignity cost? What's the price of a child's dignity? What's the price of a child's self-confidence?"

In Kevin's ideal world, there would be a bill of rights for child influencers that would include (and I'm quoting from a text he sent me here):

> (1) the documented right to say "no to being filmed"; (2) a "right to anonymity" in which a child can choose not to have his/her likeness posted on public platforms. A child should not have to wait until the age of eighteen to demand that his/her [name or likeness] be removed from social media; (3) the right to know and also approve content creative proposals that involve them, especially for brand deals; (4) the right (or requirement) to third-party representation or agency that is not the child's parent; (5) the right to join a child influencers' union that represents all child influencers on social media; (6) laws that prohibit publicly sharing any personal details about a child prior to a certain age, and only with informed consent from the child and his/her agent after that age (this would effectively kill family vlogging); (7) the right to know how social media platforms are programming their content suggestion algorithms to promote content with your [name or likeness] in it.

In March of 2025, Utah Governor Spencer Cox signed HB 322 into law, making Utah the fourth state to legislate the profits of influencer kids. Under the Utah law, parents who make at least $150,000 per calendar year and feature their minor children in monetized or sponsored social media content must maintain a monthly record of that content for at least two years; they must also establish a trust into which a percentage of any earnings must be transferred to be held until the child reaches the age of eighteen. The formula for the percentage of earnings involves a complex calculation between the minor's earnings, the paid minutes featuring the qualified minor, and the preceding month's income from social media, among other factors. In my reporting on this legislation, Shari Franke exclusively told me for *Teen Vogue*, "I have felt such a calling to help and protect kids who went through the same experiences that I did. I am proud of the Utah legislature for taking this step in protecting child influencers. However, my work is only beginning, and I will continue to be outspoken on putting an end to family vlogging."

In July of 2025, Virginia became the fifth state to pass legislation related to child influencers when HB2401 took effect. Similarly to the Utah law, Virginia's law requires that if at least 30 percent of a content creator's compensated video content produced within a 30-day period includes the likeness, name, or photograph of a child who is under the age of sixteeen, and the content is compensated at a rate of $.10 or more per view, a percentage of the earnings must be put into a trust for the child.

Bills to protect child influencers have been introduced in several other states, including Ohio, New York, Washington, and Montana. Jess Maddox, an associate professor at the University of Alabama who studies social media platforms and pop culture, said we're in a moment of "groundswell support" for this type of legislation. "Everyone is starting to realize using kids as content is not okay. And it's one of those things I think we'll look back on in definitely twenty—hopefully, earlier than that—years, and be like, this is not something we should have ever been doing. Like, this is exploitation."

I wonder about the laws that have been passed. Are they meaningfully enforceable or largely symbolic? The exact nature of family influencing is what makes it so difficult to legislate. A child influencer's work unfolds as their life does. How can labor standards be applied to work that is done inside the home as part of daily life? It can even be difficult to quantify that effort and figure out just how much work a child influencer is doing, especially when the boundary between their home life and their work life can be so tenuous. If a child is being filmed as they go about their daily life, is that work? If the end product is a 30-second video, how much work was behind that? "No one really knows how much time goes into [content creation]. It's hard to know what meaningful and enforceable legislation would look like," says Francis Rees, a lecturer in law at the University of Essex and the coordinator for the Child Influencer Project.

The labor of a child influencer is so different than that of child actors, to whom they are often compared. A child actor goes to a film set to do their work, creating a clearer delineation between their work lives and their home lives. On that set, the child actor is surrounded by production assistants and camera people who are clearly working; there are teachers who have to spend a certain number of hours each day on the child actor's schooling. There are regulations around the number of hours the child can work per day and what time their work must end. For the child influencer, there is none of that—there is only their parents and the limits of the laws that have been passed in an attempt to protect the children at the center of monetized content. "The problem with these laws is that you can't regulate how people treat their children in their living rooms or their backyards the way that you can when you have a permit and you're on set," says Karen North, a professor of communication at the University of Southern California.

That can be especially difficult in the United States, which tends to prioritize parents' rights over the rights of children, as noted in Annie Dunn's paper "The Kids Aren't Alright: Creating Greater Protections for the Children of Family Vloggers." "In the context of family vloggers, it seems likely that legislatures will uphold the longstanding tradition of parental discretion in this country," writes Dunn. "If a parent has decided that it is in the best interests of their child to appear in their social media content, any privacy interests a child might have do not supersede the right of the parent to make that decision." Put simply: The United States places the utmost value on the ability of parents to make choices for their children. If a parent decides to be a family vlogger and make their child a kid influencer, that's their choice.

All five of the laws that have been passed related to child influencers include a formula of how often a minor child appears in monetized content. For example: Illinois's law is triggered if a minor appears in 30 percent of a creator's monetized content within a 30-day period. That includes the "likeness, name, or photograph" of the minor. Though this seems relatively straightforward, it's actually not, says Kameron Monet, a New York lawyer and content creator. "How do we measure the level of involvement the child had? That's going to be a big question," Monet says. "There are a lot of nuances that aren't answered." If a child is in the background of a video, does that count as an appearance? If the back of their head is showing but their face is hidden, does that count? Those answers will come, she says, when a child influencer sues their parents under the law and the courts are forced to reckon with the questions raised. Even as a lawyer who counsels creators, Monet finds the formulas of these laws complicated. Let's take a look at Utah's.

> (5) (a) If a content creator's content had minor content earnings in the previous month, a content creator shall use the formula E = (A/T) * (Q/S) * (M/2) or the formula E = (A/T) * (1/X) * (M/2) to determine the qualifying minor's earnings to transfer to each qualifying minor where:
>
> (i) E = a qualifying minor's earnings;
> (ii) A = all paid minutes featuring any qualifying minor;
> (iii) T = total paid minutes;
> (iv) Q = paid minutes featuring the qualifying minor;
> (v) S = the sum of paid minutes for all qualifying minors;
> (vi) M = the preceding month's income from social media; and
> (vii) X = the total number of qualifying minors; and

(b) (i) subject to Subsection (5)(b)(ii), transfer the minor's earnings described in Subsection (5)(a)(i) directly to the qualifying minor; or

(ii) after the qualifying minor's parent or guardian establishes a trust as described in Subsection (3)(c), transfer the qualifying minor's earnings calculated to each qualifying minor's trust.

How is the average content creator—even if they have the best intentions—supposed to ensure that they're in compliance with this law? It seems needlessly complicated, and it's not the only one.

The Minnesota law is activated when a minor appears in 30 percent of monetized content in which at least $.01 is earned per view. This sounds simple on the surface, but when I think of this as someone who posts monetized content myself, it's really complex. Let's look at an example. I recently posted a TikTok that got 23,000 views. The estimated rewards for that video are currently $13.38. So, if you divide $13.38 by 23,000, you get $.00582 per view—less than the $0.01/view threshold that triggers the Minnesota law. So does a content creator have to go through every single piece of monetized content and do these calculations? And then, if they reach the $.01/view threshold, do they then have to calculate how much of the content their child is featured in? If they hit that 30 percent threshold, then they would theoretically have to calculate what percentage of the money should be saved. You would need a forensic accountant to make sure every family vlogger and mom influencer is in compliance with these laws.

Monet, the lawyer and creator, says the convoluted nature of the laws is a direct result of who's writing them. "A lot of the decision-makers—lawyers, legislators, et cetera—don't understand practically how influencing works," she says. "It's the Wild Wild West because law is clueless." Eric Farber, a California-based lawyer and CEO of

Creators Legal, which provides legal representation for creators, says the formulas make no sense. "How do they come up with this stuff?" he says. The protections the laws provide, Farber says, are only as good as the parents, managers, and lawyers who are abiding by them in the interest of the child, though the legislation does create a legal framework for a child influencer to take legal action against their parents if a trust hasn't been established.

Both the Minnesota law and the Utah law feature components related to the deletion of content. Under the Minnesota law, if a person who is thirteen years or older requests the deletion of content featuring their likeness, the content "must be deleted." One Minnesota lawyer, who requested to remain anonymous because they hadn't been cleared to speak with me, said she wonders if this statute will affect the viability of brand deals featuring children. If brands are worried that kids may eventually request the deletion of sponsored content, the brands may not want to include the children at all.

Under the Utah law, an individual cannot request the deletion of content until they reach the age of eighteen. And unlike the Minnesota law, the Utah law doesn't purport to guarantee the deletion of content—it only provides that the request can be made.

When I first read this part of the Utah legislation, I have to admit I was confused. Don't we all technically have the right to *request* deletion of content? After all, this law doesn't require that the parent acquiesce to the request for deletion—it just gives the featured minor child the right to request it. How does that change anything? Monet explains it this way: If a child (or adult who was featured in content as a child) requests that their parents delete content and their parents refuse the request, this law establishes a framework for them to sue their parents for the refusal. In our conversation, I ask Monet what she thinks would happen if an influencer parent did refuse to delete content featuring their minor child. Could the platforms themselves be pushed to delete the content without the consent of the parent who posted it? Monet says this is one of the details that isn't spelled out in this legislation, but she doubts it. "The right of deletion is between the parent and the child," she says.

According to Ari Cohn, an Illinois-based lawyer who specializes in the First Amendment, the law is written so broadly that *any* content posted of a minor could be subject to a deletion request. This could include journalistic articles featuring a minor or even TikToks made by a fellow minor. "It's unconstitutional," Cohn says of the statute. "Speech doesn't just become unprotected because it features a minor." The intention of legislators passing these laws might be pure, Cohn says, but the free speech implications are troubling. I also think about the reality of deletion for a child influencer. In the research for this book, there were many times where I found myself watching content that had been deleted by the original family vlogger or mom influencer who had posted it. But because they have such large fan bases, the content has been screenshotted and recorded and reposted, and I could find it easily. How would a child influencer, or the law, handle the internet hydra of reuploaded, unregulated content?

Maureen Flatley, a child exploitation and safety expert, is dubious about the efficacy of child influencer legislation. She points to the case of Ruby Franke. "She was the one exploiting the kids," Flatley says. Under the Utah law, Franke would have been tasked with keeping meticulous records and establishing trust funds for her children, which Flatley doubts would have occurred. "It's almost like the kids need to have a third-party trustee or something," she says. "When you start to say that the parent, who in this case was overtly abusive, should be the child's trustee—well, that probably won't work." Taylor Lorenz, the internet culture reporter, shares Flatley's concern. "Under the Utah law, parenting content creators are explicitly allowed to be trustees for their children's trusts. This gives these allegedly abusive parents even more control over their child's finances," Lorenz wrote in a newsletter.

She also sees the laws as painting the practice of child influencing with a broad brush—as inherently exploitative. "For some reason, people cannot understand the concept of something being labor without being exploitative," Lorenz says. She considers the backlash against mom influencers and family vloggers to be misplaced. "Everybody wants to critique the mom influencers and not the tech companies that are creating these incentives to begin with," she says. I understand what she means; parents wouldn't be making their kids into influencers if they weren't incentivized to do so. But I don't believe that the ultimate responsibility lies with tech companies. It's a parent's job to protect their child. Tech companies are, after all, companies—they exist to make money, not protect children. That's what I find so confusing about the insistence that tech companies like Meta, Google, and Bytedance take steps to protect child influencers. Why would they? Child influencers make profit hand over fist for these corporations. It's not in their interest to institute protections. In my view, the job of protection lies with parents, though of course not every parent can be trusted equally. In a research paper titled "Famous at Five: Risk Assessing Digital Child Labour," lawyer and researcher Francis Rees suggests the possibility of a digital child content commissioner who can act as a third-party advocate. The commissioner, Rees writes, "could act as an advocate for the best interest of the child, working with parents and educating them on their responsibilities, as well as supporting them with issues such as contract negotiation, data security, insurance advice, and risk assessments."

Though advocates cheer every time another state passes a law protecting the profits of influencer children, Sam, the onetime highly successful family vlogger, is less optimistic about the idea of legislative solutions being focused on the money these channels and accounts are making. Remember when Sam told me that they once bribed their children $1,000 each to participate in a video? The issue isn't that influencer kids aren't being paid, Sam says. Lots of them are. Consider them incentives or bribes or just the profit they deserve, but Sam says most influencer kids are being paid, and when legislative solutions focus on the monetary aspect of child influencer work, Sam worries that they're legitimizing the institutions of family vlogging and child influencing. "These laws just state that the parents need to financially compensate the children for the work. But what I want people to understand is that's been happening the entire time, and it's not going to change anything. It's those incentives that are being used as bribes to manipulate children into doing, saying, participating in things that they otherwise probably wouldn't do," Sam says.

If the issue of compensation is not the biggest issue to Sam, what is the biggest issue? "The emotional and psychological trauma and damage that comes along with posting somebody's life, somebody's image, somebody's character for the entire public to see and come to know," they say. And none of that, Sam worries, is going to be changed by these laws.

Conclusion

When I started writing this book, I thought I knew what I would find—the story of influencer children and the parents propping them up didn't seem that complicated. But through my conversations with experts, with influencers, and with the kids at the center of this issue, I'm finding that I have much more empathy for and understanding of the mom influencers and family vloggers I've studied than I could have ever foreseen.

I think of those first mom bloggers who changed the way we talked about pregnancy and the postpartum period and motherhood. When I nurse my baby and scour the internet for personal essays on the exact part of motherhood that's sending me to the brink at that particular moment, I have those mommy bloggers to thank for the frank writing and conversations I unearth in my searches. Those first blogs were so tender—the way they were, in the beginning, writing only to each other and to themselves, trying to fashion a path to light their way out of the disorienting and disillusioning days of early motherhood. I imagine them putting their babies to sleep before crouching over their laptops, clad in milk-stained shirts and yesterday's makeup, typing and typing and hoping. How could they have known how their writing would morph into an entire industry that would sprout around their work?

I think of the mom influencers and family vloggers of today who, in viral videos, attend to the daily work of being a parent. In front of a camera, they unload the dishwasher and comb their kids' hair and toss pasta into a boiling pot of water for dinner. After I put my own daughter to sleep and wonder how it's possible to be this tired and still function, I turn to my phone and watch momfluencers go through the same motions I've just gone through, identical but completely different.

This book has been my attempt to insert nuance into the conversation around mom influencers and family vloggers and the work they do (because yes, it is work). I don't think it's productive to say that every parent who features their child in monetized content is immoral and evil and moneygrubbing. It's just not true. The truth is that there are some children who love being YouTube and TikTok stars and some for whom it's a nightmare. There are some whose parents ignore them in their pursuit of viral fame and some for whom the chasing of subscribers and views is a familial undertaking. There are parents who overwork their children and post wildly personal updates about them, and there are parents who have appropriate boundaries and try to make their kids' lives as normal as possible even as they become internet celebrities.

The flattening of the narrative around child influencers does nothing to protect them. If your purported interest is to shield the kids at the center of this multibillion-dollar industry—their privacy and their profits and their informed consent—you have to be honest about what it's like to be part of it. For some kids, it's heaven. For others, it's hell. And we owe it to these kids to listen to all the stories they're willing to tell. There is both Rachel, who told me she wishes her mom would stop filming her constantly and just have a real conversation with her, and Zac, who relishes his YouTube celebrity status and hopes to extend it into his adult life. And in a way, child influencer stardom is just another choice parents make for their kids. The job of every kid, as they grow up, is to reckon with the choices their parents have made.

(Though, to be fair, most kids don't have to do all of this—the growing and the reckoning—in front of an audience of millions.)

Viral fame and the money that follows it can change the lives of the families who earn it—some of whom have few paths out of their financial circumstances. Alexandra, after going viral on TikTok, was able to move her family out of government-subsidized, low-income housing and into a house she bought. MariClare, who gave birth at just fifteen, was able to secure an apartment for herself and her daughter. Bethanie is the breadwinner for her five children as a high school graduate. In these situations, I can't help but think that the loss of privacy may be worth the rewards and security of being an influencer—and being an influencer's child. And the truth is, it is up to the parents. They weigh their children's inability to give informed consent, the potential loss of privacy, and the intrusion of cameras and filming against their own ability as a parent to give their children a financially secure life full of opportunity. I can see how the scales tip.

I think it's dishonest for people to pretend they would never make that trade, especially in the United States, a country where research has shown that working mothers, on average, earn less than working fathers in *every* state, there is no federally mandated maternity leave (with most mothers returning to work within *ten weeks* of giving birth), and mothers are less likely to be hired or promoted compared to people without children. With so few career choices that are compatible with the demands of pregnancy and motherhood, is it any wonder that influencing and vlogging becomes so attractive? It's a career where your identity as a mother is not only accepted but considered a bonus. It's a job where you don't have to choose between promotion and being able to leave at a reasonable time to pick your kids up from school. What other job can you think of like that?

But also: There are things I've seen during the reporting of this book that I won't stop thinking about for a long time. I've seen videos of kids attending funerals of their grandparents and crying over their absent fathers and getting the sex talk and hiding from their parents who are badgering them about their crushes. I've seen photo after photo of sick, sad kids. It's too much. I don't need to know your child's potty-training routine. I think it's borderline insane to film your child's tantrums and upload them to the internet. Some things should be kept private, even when you know how viral the video of them would go (and how much money would land in your pocket). Even if you choose a career that hinges on sharing your life—and the lives of your children—there should be boundaries.

I think back to the early mom bloggers and how much of their work focused on their experience of motherhood rather than the lives of their children. There has to be a way to have both: to be able to write and make content about the life-altering, dizzying experience of becoming a parent while also protecting your child's privacy. (Even as I mention my daughter in passing in this book, I wonder: *Is this too much?*) I think about the moms who post content without showing their children's faces or sharing their names. There is some measure of privacy there for the kids. There is autonomy and a career for the content creators. And there is solace too for moms like me who are fumbling around for a light in a dark room.

I'm talking in circles. Let me clarify. Here is where I have landed, after hundreds of interviews and thousands of hours of researching and writing. The money is eye-watering; the fame is intoxicating; the possibility for economic freedom is dizzying. I know all of that. And still, I would never—ever, not for a second—make the trade.

Acknowledgments

I sold this book when I was seven months pregnant and turned in the first draft at seven months postpartum, which means that first and foremost I want to thank the people who helped me attempt to balance pregnancy and early motherhood with writing.

First: to my husband, Alexander, who makes me laugh more than anyone in the world. You make hard things easier and good things more beautiful. We have the best girl together. I love you forever.

To my mother, Drita, and my mother-in-law, Birgit, both of whom never hesitate to jump on a plane when I need them. You held my girl for countless hours while I chipped away at this book; you held me, too. If I'm half the mother the two of you are, my daughter will be lucky.

To my sister, Qendresa, who takes care of me in a way that allows me to take care of my daughter. Thank you for the countless dinners you've made us (and everything else). I'll owe you always. And to Ross, thank you for making the family compound a real thing.

To Annie, our incredible nanny. Thank you for loving my girl. To be able to listen to the two of you laugh while I worked on this book was a beautiful gift.

Thank you to my father, Rifat, who insists his children are capable of everything they can dream of and whose belief in that has changed our lives. Eye on the flag.

And again, and always: to my mother, Drita. You are all of our North Star.

To my siblings: Kalterina, Qendresa, Kushtrim, and Lulejeta. I wouldn't want to be anything other than a Latifi sibling.

To my nieces and nephews: Zana, Adelina, Norik, Ellis, and Eleanora. My babies before my baby, and still.

To Neeaz, for taking me seriously as a writer since that poem I wrote you in the third grade. I'm so grateful our child selves chose each other.

To Lily, my comrade in the mind-bending transition into motherhood. I don't know where I would be without our constant text conversation, which hasn't ended since our daughters were born.

To my father-in-law, Edwin, who reminds me that if you don't A-S-K, you don't G-E-T.

To MacKinley, who dove into the weeds of research. Thank you for joining the team.

To Mina, the agent of my dreams. Thank you for answering every frantic text I send your way. This is just the beginning.

To Rebecca, the editor without whom this book would barely exist. You made every sentence better. To Aimée, Taylor, Max, and the rest of the wonderful team at Gallery: thank you.

And finally: thank you to anyone who has taken the time to read my stories and watch my videos and engage with my work. You helped make this possible.

Notes

Introduction

ix *In a review / "comfortably ordinary"*: Anne Roiphe, "An American Family," *New York Times*, February 18, 1973, https://www.nytimes.com/1973/02/18/archives/things-are-keen-but-could-be-keener-an-american-family-an-american.html.

x *Family vlogging and*: Darian Woods and Wailin Wong, "The Promise and Perils of the Multi-Billion Dollar Influencer Industry," NPR, May 5, 2023, https://www.npr.org/2023/05/05/1174242181/the-promise-and-perils-of-the-multi-billion-dollar-influencer-industry.

1. Dear Diary, Meet the Algorithm:

How Mommy Bloggers Paved the Way

2 *One of those women*: Catherine Connors, *Her Bad Mother* (blog), 2006, https://herbadmother.com/.

2 *In one of*: Catherine Connors, "How to Raise an Asshole," *Her Bad Mother* (blog), March 3, 2020, https://herbadmother.com/2020/03/how-to-raise-an-asshole/.

4 *Rebecca Woolf / from 2005 to 2017*: Rebecca Woolf, *Girl's Gone Child* (blog), 2005, http://www.girlsgonechild.net/.

7 *When I call*: Shannon Bird (@Birdalamode), Instagram account, https://www.instagram.com/birdalamode/.

12 *In a piece*: Lydia Kiesling, "What Happened to Natalie Jean, a.k.a. Nat the Fat Rat?," *The Cut*, June 20, 2018, https://www.thecut.com/2018/06/what-happened-to-natalie-jean-nat-the-fat-rat.html.

12 *She's on Instagram*: Natalie Lovin (@heynataliejean), Instagram account, https://www.instagram.com/heynataliejean/.

13 *Some of her / like Jo Goddard*: Joanna Goddard (@CupofJo), Instagram account, https://www.instagram.com/cupofjo/.

15 *In her book*: Taylor Lorenz, *Extremely Online: The Untold Story of Fame, Influence, and Power on the Internet* (New York: Simon & Schuster, 2023), 22–27.

2. A Filtered Life:

Fame, Family, and Clickbait

17 *"Let's make a period kit"*: Aubree Jones (@WhatAboutAub), Instagram account, https://www.instagram.com/whataboutaub/; Aubree Jones (@WhatAboutAub), "#ad I love that @ubykotex makes pads specifically for teens that I could add to this period kit!," Instagram post, December 4, 2024, https://www.instagram.com/p/DDK6NFByQoF/.

22 *In the research / "These interactions require"*: Francis Rees, "Famous at Five: Risk Assessing Digital Child Labour," *Information & Communications Technology Law* (January 2025): 15, https://doi.org/10.1080/13600834.2025.2452699.

27 *When I first*: Jamie Otis Hehner (@jamienotis), Instagram account, https://www.instagram.com/jamienotis/.

27 *One of her pinned*: Jamie Otis Hehner (@jamienotis), "This was just moments before my baby became unresponsive, stopped breathing and his lips turned blue," Instagram post, June 1, 2022, https://www.instagram.com/p/CeRzYvkAPn_/.

29 *Take Julie Jeppson*: Julie Jeppson (@TheBigFamilyJewels), YouTube account, https://www.youtube.com/channel/UCHI_BZNa02ETMuPCbiHShag.

37 *"How does a"*: Robyn Koslowitz, "The Slippery Slope of Sharenting," *Psychology Today*, February 11, 2025, https://www.psychologytoday.com/us/blog/targeted-parenting/202502/the-slippery-slope-of-sharenting.

37 *This paradox was*: Emma Beuckels et al., "Work It Baby! A Survey Study to Investigate the Role of Underaged Children and Privacy Management Strategies Within Parent Influencer Content," *New Media & Society* 27, no. 6 (January 2024): 3081–101, https://doi.org/10.1177/14614448231218992.

3. Privacy for Sale:

Monetizing Everyday Intimacies

39 *The Family Fun Pack YouTube channel*: Kristine (@FamilyFunPack), YouTube account, https://www.youtube.com/user/familyfunpack.

39 *There are eight*: Kristine, "Twins Putting Themselves to Bed," Family Fun Pack, October 5, 2011, YouTube video, 1:13, https://www.youtube.com/watch?v=fJ7qFXneLpk.

42 *One of his*: Kristine, "Eating Only ONE Color of Food for 24 Hours!!!," Family Fun Pack, October 12, 2019, YouTube video, 27:40, https://www.youtube.com/watch?v=RYwMOKT484k.

42 *A few months*: Kristine, "Potty Train Your Child in 1 Week! 10 Tips—Mom of 8," Family Fun Pack, June 7, 2024, YouTube video, 20:48, https://www.youtube.com/watch?app=desktop&v=tVHF0iOK610.

44 *They have fans*: Matthew Townsend, "Influencer Nation: 86% of Young Americans Want to Become One," Bloomberg News, November 5, 2019, https://www.bloomberg.com/news/articles/2019-11-05/becoming-an-influencer-embraced-by-86-of-young-americans.

45 *While studying for*: Bridie Hamilton, "Anything for Views Parenting: Framing Privacy, Ethics, and Norms for Children of Influencers on YouTube," master's thesis, Duke University (2023): 17–32, https://hdl.handle.net/10161/27884.

46 *Consider "Our First"*: Cole and Savannah LaBrant (@ColeandSav), YouTube account, https://www.youtube.com/@ColeAndSav; Cole and Savannah LaBrant, "Our First Pet Died," The Labrant Fam, August 4, 2018, YouTube video, 10:29, https://www.youtube.com/watch?v=In01Q8jFlJw&ab_channel=TheLaBrantFam.

46 *Take, for example, Christian*: Caila Burrello (@cailaquinn), Instagram account, https://www.instagram.com/cailaquinn/.

47 *Caitlin Nichols, an influencer*: Caitlin Nichols (@555mph.a.triplet.story), TikTok account, https://www.tiktok.com/@555mph.a.triplet.story.

48 *Take Brooklyn and Bailey*: Brooklyn and Bailey McKnight (@brooklynandbailey), Instagram account, https://www.instagram.com/brooklynandbailey/; Brooklyn and Bailey McKnight (@brooklynandbailey), TikTok account, https://www.tiktok.com/@brooklynandbailey; Brooklyn and Bailey

McKnight (@brooklynandbailey), YouTube account, https://www.youtube.com/user/BrooklynAndBailey.

48 *Mindy McKnight, who*: Brooklyn and Bailey McKnight (@brooklynandbailey), TikTok account, https://www.tiktok.com/@brooklynandbailey; Mindy McKnight (@CuteGirlsHairstyles), YouTube account, https://www.youtube.com/channel/UC2LgZ_4GzSFQS-3a87_Jc6w.

48 *By the time Brooklyn*: Madeline Stone and Lauren Browning, "13 YouTube Stars Who Should Be on Your Radar," *Business Insider*, February 23, 2015, https://www.businessinsider.com/up-and-coming-youtube-stars-to-follow-2015-2.

49 *In 2026, they*: ITK Skincare (@stayitk), Instagram account, https://www.instagram.com/stayitk/; Lash Next Door (@lashnextdoor), Instagram account, https://www.instagram.com/lashnextdoor/.

49 *Their first video*: Brooklyn and Bailey McKnight, "Welcome to BrooklynAndBailey! | Intro Video," Brooklyn and Bailey, March 17, 2013, YouTube video, 5:36, https://www.youtube.com/watch?v=Dk_H_GyEkKA.

49 *Two years later*: Brooklyn and Bailey McKnight, "Get Ready with Us: HOMECOMING 2015," Brooklyn and Bailey, September 30, 2015, YouTube video, 9:19, https://www.youtube.com/watch?v=gFgHpWaHHw8.

49 *They debate where*: Brooklyn and Bailey McKnight (@brooklynandbailey), "Choosing Where to Go to College Is One of the Hardest Decisions You'll Ever Make," Instagram post, March 31, 2019, https://www.instagram.com/p/BvrsnRvHKSg/.

50 *Although she makes*: Brooklyn and Bailey McKnight, "Baby Shower Haul (First Time Mom)," Brooklyn and Bailey, October 23, 2024, YouTube video, 12:10, https://www.youtube.com/watch?v=LgffGa1XOSQ; Brooklyn and Bailey McKnight, "Brooklyn Had Her Baby | Unmedicated Birth *EMOTIONAL*," Brooklyn and Bailey, January 22, 2025, YouTube video, 25:25, https://www.youtube.com/watch?v=uiJfr3qGsXc.

50 *She is the*: Shay Butler (@shaytards), YouTube account, https://www.youtube.com/user/SHAYTARDS.

51 *Now, Avia has*: Avia Colette Butler (@avia_colette), TikTok account, https://www.tiktok.com/@avia_colette; Avia Colette Butler (@AviaColette), YouTube account, https://www.youtube.com/@AviaColette; Avia Colette Butler (@AviaColette), Instagram account, https://www.instagram.com/aviacolette/.

51 *In a 2023*: Avia Colette Butler (@AviaColette), "I wouldn't of [sic] Had It Any Other Way!," TikTok video, December 12, 2023, https://www.tiktok.com/@avia_colette/video/7312201772305599775.

52 *Everleigh LaBrant is*: Everleigh LaBrant (@everleighrose), Instagram account, https://www.instagram.com/everleighrose/.

52 *There is an entire*: LabrantFamSnark (r/LaBrantFamSnark), Reddit page, https://www.reddit.com/r/LaBrantFamSnark/.

53 *In a YouTube*: Cole and Savannah LaBrant, "How We Met | Cole & Savannah," The Labrant Fam, September 22, 2016, YouTube video, 9:48, https://www.youtube.com/watch?v=qonmB9pWpOY.

54 *On YouTube, Posie's*: Cole and Savannah LaBrant, "Meeting Our Baby Girl for the First Time. (Live Birth)," The Labrant Fam, December 31, 2018, YouTube video, 11:59, https://www.youtube.com/watch?v=pGSiBhR3eZA.

54 *After Beckham's birth*: Cole and Savannah LaBrant, "We're Done Having Kids. (Coles [sic] Having Surgery)," The Labrant Fam, July 1, 2024, YouTube video, 25:07, https://www.youtube.com/watch?v=jwSS5NLgQP4.

55 *In 2022, Everleigh's*: Savannah Rose LaBrant (@sav.labrant), "Our hearts are incredibly heavy as we process the loss of Everleigh's dad, Tommy," Instagram post, September 14, 2022, https://www.instagram.com/p/Cif4TE0P7Bp/.

56 *Which isn't that rare*: Stacey Steinberg, "Sharenting: Children's Privacy in the Age of Social Media," *Emory Law Journal* 66, no. 4 (2017), https://scholarlycommons.law.emory.edu/elj/vol66/iss4/2.

57 *Take Ryan Kaji*: Ryan Kaji (@RyansWorld), YouTube account, https://www.youtube.com/channel/UChGJGhZ9SOOHvBB0Y4DOO_w.

58 *Ryan became the most popular*: E. J. Dickson, "He Got Famous at 3. How Does a YouTube Superstar Grow Up?," *Rolling Stone*, August 15, 2024, https://www.rollingstone.com/culture/culture-features/ryans-world-ryan-kaji-youtube-1235078164/.

58 *In his first*: Ryan Kaji, "Kid playing with toys Lego Duplo Number Train," Ryan's World, March 17, 2015, YouTube video, 15:13, https://www.youtube.com/watch?v=jVwSJ9q3kOc.

58 *Now twelve years old:* Jay Caspian King, "The Boy King of YouTube," *New York Times,* January 5, 2022, https://www.nytimes.com/2022/01/05/magazine/ryan-kaji-youtube.html.

58 *That's something Bekah*: Bekah Martinez (@bekah), Instagram account, https://www.instagram.com/bekah.

4. When the Cameras Don't Stop:

Meltdowns, Mistakes, and Crossing Moral Lines

64 *In an interview*: Rachel Paula Abrahamson, "Family YouTuber Deletes Account After Criticism Over Video Coaching Son to Cry," *Today Parents*, September 14, 2021, https://www.today.com/parents/jordan-cheyenne-speaks-out-about-youtube-video-son-crying-t231055.

64 *"It's so wrong"*: Jordan Cheyenne (@jordancheyenne), YouTube account, https://www.youtube.com/jordancheyenne.

64 Rolling Stone *published*: C. T. Jones, "Fans Built Her an Internet Empire. Now They're Tearing It Down," *Rolling Stone*, June 21, 2023, https://www.rollingstone.com/culture/culture-features/colleen-ballinger-miranda-sings-youtube-fans-allegations-1234774947/.

64 *Though she has*: Colleen Ballinger (@colleen), Instagram account, https://www.instagram.com/colleen/.

66 *In a viral*: Caitlin Moscatello, "Why Did These YouTubers Give Away Their Son?" *The Cut*, January 15, 2025, https://www.thecut.com/2025/01/youtube-myka-james-stauffer-huxley-adoption.html.

66 *In the story from* The Cut: Caitlin Moscatello, "Why Did These YouTubers Give Away Their Son?" *The Cut*, January 15, 2025.

68 *Myka hasn't posted*: Myka Stauffer (@mykastauffer), Instagram account, https://www.instagram.com/mykastauffer/.

68 *Her husband, James*: James Stauffer (@StaufferGarage), YouTube account, https://www.youtube.com/channel/UC6f1N1Cg__5Ex0aWBUZeYGg.

69 *Their content included prank*: Piper Rockelle, "Randomly Crying Throughout the Day Prank!!," Piper Rockelle, February 27, 2022, YouTube video, 14:07, https://www.youtube.com/watch?v=O6Z2huwU9q8.

69 *and challenge videos*: Piper Rockelle, "Last to Leave the Trampoline Wins $10,000 Challenge **Gone Wrong** | Piper Rockelle," Piper Rockelle, September 7, 2019, YouTube video, 34:35, https://www.youtube.com/watch?v=DckpTdNvuHo.

69 *The claims in the lawsuit*: Dhillon Law Group Inc., "YouTube Influencers v. Piper Rockelle Inc.," January 12, 2022, https://www.dhillonlaw.com/lawsuits/youtube-influencers-v-piper-rockelle-inc/.

69 *Though the defendants*: Angela Yang, "YouTube Mom's Child Abuse Scandal Ends in $1.85 Million Settlement," NBC News, October 4, 2024, https://www.nbcnews.com/tech/piper-rockelle-mom-youtube-settlement-deal-rcna174615.

70 *In the aftermath*: Fortesa Latifi, "Piper Rockelle Has a Lot to Say About That Netflix Documentary," *Rolling Stone*, May 1, 2025, https://www.rollingstone.com/culture/culture-features/piper-rockelle-interview-bad-influence-netflix-1235328719/.

71 *Perhaps the most*: Kelsey Ables and Kim Bellware, "What to Know About Ruby Franke, Parenting YouTuber Charged with Child Abuse," *Washington Post*, September 1, 2023, https://www.washingtonpost.com/nation/2023/09/01/ruby-franke-youtube-8passengers-child-abuse/.

72 *The Santa Clara–Ivins*: Jaron Studley, "Santa Clara–Ivins Public Safety Department Press Release, Incident 23SCI4142, Child Abuse," press release, August 31, 2023, https://media.washco.utah.gov/franke-hildebrandt/Documents/Santa%20Clara-Ivins%20PD%20Press%20Release.pdf.

72 *She was sentenced*: C. T. Jones, "Mommy Vlogger Ruby Franke Will Spend Up to 60 Years in Prison for Child Abuse," *Rolling Stone*, February 20, 2024, https://www.rollingstone.com/culture/culture-news/ruby-franke-child-abuse-true-crime-sentenced-1234971427/.

75 *As I write*: Hannah Bhiatt (@hannah_bhiatt), TikTok account, https://www.tiktok.com/@hannah_bhiatt.

75 *Hannah first gained*: Rosie Colosi, "After a Story About '17 Diapers,' Moms Are Sharing Their Own Tales of Feeling Overwhelmed," *Today*, October 16, 2024, https://www.today.com/parents/moms/17-diapers-mom-postpartum-rcna175282.

75 *Nurse Hannah told*: Angela Andaloro, "New Mom Speaks Out After Video of Her Finding 17 Dirty Diapers in Her Home Goes Viral," *People*, October 15, 2024, https://people.com/mom-responds-to-backlash-after-admitting-there-were-17-dirty-diapers-in-her-home-exclusive-8728111.

77 *The Ogden Police Department*: Stephanie Fasano, Mack Muldofsky, and Doc Louallen, "Controversy Surrounding TikTok Influencer Hannah Hiatt Shows Potential Consequences of 'Sharenting' Online," ABC News, December 16, 2024, https://abcnews.go.com/US/controversy-surrounding-tiktok-influencer-hannah-hiatt-shows-potential/story?id=116724604.

5. Filters, Feeds, and First Steps:

The TikTok Teen Mom Hustle

79 *Every day, Brooke*: Brooke Morton (@brookexaloura), TikTok account, https://www.tiktok.com/@brookexaloura.

79 *There's sixteen-year-old Kylie*: Kylie (@the.noelanik), TikTok account, https://www.tiktok.com/@the.noelanik.

79 *In the bio*: Kylie (@the.noelanik_fam), TikTok account, https://www.tiktok.com/@the.noelanik_fam.

80 *MariClare Maclamroc, who*: MariClare MacLamroc (@_mariclaremaclamroc_), TikTok account, https://www.tiktok.com/@_mariclaremaclamroc_.

80 *Research shows that*: Jennifer Manlove and Hannah Lantos, "Data Point: Half of 20- to 29-Year-Old Women Who Gave Birth in Their Teens Have a High School Diploma," Child Trends, January 11, 2018, https://www.childtrends.org/publications/half-20-29-year-old-women-gave-birth-teens-high-school-diploma; Aleksandra Jakubowski et al., "Unwinding the Tangle of Adolescent Pregnancy and Socio-economic Functioning: Leveraging Administrative Data from Manitoba, Canada," *BMC Pregnancy and Childbirth* 23, no. 140 (March 2023), https://doi.org/10.1186/s12884-023-05443-6; Sheree J. Gibb et al., "Early Motherhood and Long-term Economic Outcomes: Findings from a 30-Year Longitudinal Study," *Journal of Research on Adolescence* 25, no. 1 (March 2014): 163–72, https://doi.org/10.1111/jora.12122.

82 *It was 2013*: R. Lyle Cooper et al., "Modeling Dynamics of Fatal Opioid Overdose by State and Across Time." *Preventive Medicine Reports* 20, (August 2020), https://doi.org/10.1016/j.pmedr.2020.101184.

82 *The state would*: Dietrich Knauth, "West Virginia Cities Reach $400 Mln Opioid Distributor Settlement," Reuters, August 1, 2022, https://www.reuters.com/business/healthcare-pharmaceuticals/west-virginia-cities-reach-400-mln-opioid-distributor-deal-2022-08-01/.

87 *I can understand the concern*: Centers for Disease Control and Prevention, US Department of Health and Human Services, About Teen Pregnancy (web page), May 15, 2024, https://www.cdc.gov/reproductive-health/teen-pregnancy/index.html.

88 *The show really*: Melissa S. Kearney and Phillip Levine, "Media Influences on Social Outcomes: The Impact of MTV's '*16 and Pregnant*' on Teen

Childbearing," Brookings Institution, January 15, 2014, https://www.brookings.edu/articles/media-influences-on-social-outcomes-the-impact-of-mtvs-16-and-pregnant-on-teen-childbearing/.

6. Curated Childhoods:

The Parents Behind the Posts

99 *Heather Bell is*: Heather Bell (@justthebells10), TikTok account, https://www.tiktok.com/@justthebells10; Heather Bell (@justthebells10), Instagram account, https://www.instagram.com/justthebells10/; Heather Bell (@justthebells10), YouTube account, https://www.youtube.com/channel/UCECrsbh5uGFxnycSdq9XJ7w.

100 *In her thesis*: Bridie Hamilton, "Anything for Views Parenting: Framing Privacy, Ethics, and Norms for Children of Influencers on YouTube," master's thesis, Duke University (2023), https://hdl.handle.net/10161/27884.

100 *Second, she began*: Sarah Adams (@mom.uncharted), TikTok account, https://www.tiktok.com/@mom.uncharted.

102 *The Bethune family*: Jenn and Kyle Bethune (@Beingbethunes), YouTube account, https://www.youtube.com/channel/UCcpfi5cDAJ23oJ7nkwx9FXQ.

103 *The video, which*: Jenn and Kyle Bethune (@Beingbethunes), "Is This Gabby Petito's Van Caught on Youtuber's Camera? Read Description," Being Bethunes, September 19, 2021, YouTube video, 14:03, https://youtu.be/PBp3aNAGuFM?si=n8sD4VyhF7z1Ip-T.

107 *The Jurgy family*: Nellie and Bryce Jurgy (@thejurgys), YouTube account, https://www.youtube.com/TheJurgys; Nellie and Bryce Jurgy (@thejurgys), Instagram account, https://www.instagram.com/thejurgys; Nellie and Bryce Jurgy (@thejurgys), TikTok account, https://www.tiktok.com/@thejurgys.

110 *So common that*: Bridie Hamilton, "Anything for Views Parenting: Framing Privacy, Ethics, and Norms for Children of Influencers on YouTube," master's thesis, Duke University (2023), https://hdl.handle.net/10161/27884.

7. Bless This Brand:

Mormon Influencers, Trad Wives, and the Invisible Labor of Motherhood

115 *In a commencement speech*: M. Russell Ballard, "Using New Media to Support the Work of the Church," December 14, 2007, BYU—Hawaii Commencement, https://speeches.byuh.edu/commencement/using-new-media-to-support-the-work-of-the-church.

116 *Though only 2 percent*: Pew Research Center, "2023–24 US Religious Landscape Study Interactive Database," 2025, doi: 10.58094/3zs9-jc14.

116 *Utah, the state*: Sharon Omondi, "Mormon Population by State," World Atlas, November 9, 2020, https://www.worldatlas.com/articles/mormon-population-by-state.html.

118 *"The Lord has"*: Steven E. Snow, "The Sacred Duty of Record Keeping," The Church of Jesus Christ of Latter-day Saints, April 2019, https://www.churchofjesuschrist.org/study/ensign/2019/04/the-sacred-duty-of-record-keeping?lang=eng.

119 *And to achieve*: Eric W. Dolan, "Unraveling Utah's Paradox: Study on LDS Church Members Examines High Cosmetic Surgery Rates in the Highly Religious State," *PsyPost*, September 2, 2023, https://www.psypost.org/unraveling-utahs-paradox-study-on-lds-church-members-examines-high-cosmetic-surgery-rates-in-the-highly-religious-state/.

120 *"The first commandment"*: "What Does the Bible Teach About Family?" The Church of Jesus Christ of Latter-day Saints, https://www.churchofjesuschrist.org/comeuntochrist/belong/family/bible-teachings-about-family.

121 *Natalie Jean Lovin*: Natalie Jean Lovin, *Hey Natalie Jean* (blog), http://www.heynataliejean.com/.

122 *The Mormon Church plays*: Tony Semerad, "Latest Figures Are Out on LDS Church's Investment Portfolio. See How Much It's Worth," *Salt Lake Tribune*, February 21, 2025, https://www.sltrib.com/religion/2025/02/21/lds-church-fund-is-down-almost/.

125 *At the top*: Hannah Neeleman (@ballerinafarm), Instagram account, https://www.instagram.com/ballerinafarm/; Hannah Neeleman (@ballerinafarm), TikTok account, https://www.tiktok.com/@ballerinafarm; Hannah Neeleman (@theballerinafarm), YouTube account, https://www.youtube.com/@TheBallerinaFarm.

125 *In a* New York Times: Madison Malone Kircher, "She Gave Birth Two Weeks Ago. Now She's in a Beauty Pageant," *New York Times*, January 30, 2024, https://www.nytimes.com/2024/01/30/style/ballerina-farm-mrs-world-hannah-neeleman.html.

125 *In an article*: Megan Agnew, "Meet the Queen of the 'Trad Wives' (and Her Eight Children," *The Times*, July 29, 2024, https://www.thetimes.com/world/us-world/article/my-day-with-the-trad-wife-queen-and-what-i-really-thought-of-her-qmbmmhkp8.

127 *"Tradwives leveraged maternal"*: Sophia Sykes and Veronica Hopner, "Tradwives: Right-Wing Social Media Influencers," *Journal of Contemporary Ethnography* 53, no. 4 (April 2024): 453–87, https://doi.org/10.1177/08912416241246273.

128 *For young women*: Rakesh Kochhar, "The Enduring Grip of the Gender Pay Gap," Pew Research Center, March 1, 2023, https://www.pewresearch.org/social-trends/2023/03/01/the-enduring-grip-of-the-gender-pay-gap/.

131 *"It says quite"*: Lachlan Ross and Lyn Craig, "On Digital Reproductive Labor and the 'Mother Commodity,'" *Television & New Media* 24, no. 6 (October 2022): 639–55, https://doi.org/10.1177/15274764221125742.

132 *In the paper*: Erla Gjinishi, "'Mommy Meets Money': Digitized Forms of Affective Labor Among 'Mommy' Bloggers and the Biopolitical Production of 'Life' as a Marketable Commodity," master's thesis, Lund University, (2018), https://www.lunduniversity.lu.se/lup/publication/8949836.

8. From Content to Consent:

The Parents Who Changed Their Minds

135 *My name is*: Bethanie Johnson, *The Garcia Diaries* (blog), https://thegarciadiaries.com/cgi-sys/suspendedpage.cgi.

136 *Though she doesn't post*: Bethanie Johnson (@thegarciadiaries), Instagram account, https://www.instagram.com/thegarciadiaries/.

137 *One of those*: Maia Knight (@maiaknight), TikTok account, https://www.tiktok.com/@maiaknight.

138 *This is what*: Michel Walrave et al., "Mindful Sharenting: How Millennial Parents Balance Between Sharing and Protecting," *Frontiers in Psychology* 14 (July 2023), https://doi.org/10.3389/fpsyg.2023.1171611.

139 *Aspyn Ovard—who*: Aspyn Ovard (@AspynOvard), Instagram account, https://www.instagram.com/aspynovard/. Aspyn Ovard (@AspynOvard),

YouTube account, https://www.youtube.com/channel/UCxjZe0qTFXh6jGm54LFWEDw; Aspyn Ovard (@AspynOvard), TikTok account, https://www.tiktok.com/@aspynovard.

142 *In 2017*, The: Bianca Bosker, "Instamom," *The Atlantic*, March 2017, https://www.theatlantic.com/magazine/archive/2017/03/instamom/513827/.

145 *Felix, who uses*: Felix and Morgan (@couplagoofs), Instagram account, https://www.instagram.com/couplagoofs/. Felix and Morgan (@couplagoofs), TikTok account, https://www.tiktok.com/@couplagoofs.

148 *As of this*: Meghan Moore (@meghan_moore_), TikTok account, https://www.tiktok.com/@meghan_moore_.

150 *The first TikTok*: Jillian Kalbaugh (@mommeh_dearest), TikTok account, https://www.tiktok.com/@mommeh_dearest.

152 *More and more*: Fortesa Latifi, "As the Internet Gets Scarier, More Parents Keep Their Kids' Photos Offline," *Washington Post*, April 5, 2024, https://www.washingtonpost.com/parenting/2024/04/08/parenting-keeping-kids-offline/.

153 *Fans who had*: Emilie Kiser (@emiliekiser), TikTok account, https://www.tiktok.com/@emiliekiser?lang=en.. Emilie Kiser (@emiliekiser), Instagram account, https://www.instagram.com/emiliekiser/.

153 *As I reported*: Fortesa Latifi, "A Momfluencer's Son Drowned. Now Other Parents Are Rethinking How Much They Share Online," *Rolling Stone*, May 21, 2025, https://www.rollingstone.com/culture/culture-features/trigg-kiser-drowning-emilie-momfluencer-parasocial-1235344977/.

155 *Emilie Kiser returned*: Emilie Kiser (@emiliekiser), August 28, 2026. Instagram. https://www.instagram.com/p/DN520cWkUVp/.

155 *Libby Ward, @libbywardofficial*: Libby Ward (@libbywardofficial), TikTok account, https://www.tiktok.com/@libbywardofficial; Libby Ward (@libbyward), Instagram account, https://www.instagram.com/libbyward/.

157 *A fair amount*: Kathryn Jezer-Morton, "Are You a Cycle Breaker—or Just a Grown-up?," *The Cut*, November 23, 2024, https://www.thecut.com/article/are-you-a-cycle-breaker-or-just-a-grown-up.html.

9. Imaginary Friends:

Model Mothers and the Cure for Loneliness

159 *Federally mandated maternity leave*: "Paid maternity leave by country: A global comparison for buyers," Rippling.com, May 21, 2025, https://www.rippling.com/blog/maternity-leave-by-country.

160 *At two weeks*: Sarah Kliff, "1 in 4 American Moms Return to Work Within 2 Weeks of Giving Birth—Here's What It's Like," *Vox*, August 22, 2015, https://www.vox.com/2015/8/21/9188343/maternity-leave-united-states.

161 *Research backs DiMarco*: Eloise R. Germic et al., "The Impact of Instagram Mommy Blogger Content on the Perceived Self-Efficacy of Mothers," *Social Media + Society* 7, no. 3 (August 2021), https://doi.org/10.1177/20563051211104164.

161 *The draw of*: Rachelle M. Chee et al., "The Impact of Social Media Influencers on Pregnancy, Birth, and Early Parenting Experiences: A Systematic Review," *Midwifery* 120 (May 2023), https://doi.org/10.1016/j.midw.2023.103623.

163 *On Instagram, Campbell*: Campbell Puckett (@campbellhuntpuckett), Instagram account, https://www.instagram.com/campbellhuntpuckett/.

163 *Emma, a thirty-two-year-old*: Taylor Giavasis (@taylorgiavasis), Instagram account, https://www.instagram.com/taylorgiavasis/.

164 *Elizabeth, a thirty-year-old*: Val (@lovelyluckylife), Instagram account, https://www.instagram.com/lovelyluckylife/.

165 *In the glowing*: Annalee Grace (@annalee15), Instagram account, https://www.instagram.com/annalee15/; Lauren Burnham Luyendyk (@laurenluyendyk), Instagram account, https://www.instagram.com/laurenluyendyk/.

165 *On TikTok, one*: Rocklyn Belle (@rocklyn.belle), TikTok account, https://www.tiktok.com/@rocklyn.belle; Lottie Weaver (@lottie..weaver), TikTok account, https://www.tiktok.com/@lottie..weaver.

165 *I scroll and*: Tiffany and Caleb Remington (@ustheremingtons), TikTok account, https://www.tiktok.com/@ustheremingtons.

165 *I scroll again*: Julie Wise (@juliewise4), TikTok account, https://www.tiktok.com/@juliewise4.

165 *Another scroll and*: Elizabeth Kuras (@homeofizzy), TikTok account, https://www.tiktok.com/@homeofizzy.

166 *"I'm your biggest"*: Addie McCracken (@addie_mccracken123), TikTok account, https://www.tiktok.com/@addie_mccracken123; Brady Stauffer (@bradystauffer15), TikTok account, https://www.tiktok.com/@bradystauffer15.

170 *One former fan*: Ellie and Jared (@ellieandjared), YouTube account, https://www.youtube.com/channel/UCYv8VkKxvmfYIRbowQALwCw.

171 *On her new*: Karissa Collins (@thecollinskidsfamily), Instagram account, https://www.instagram.com/thecollinskidsfamily/.

10. From Love to Hate:
Former Fans, Parasocial Relationships, and the Rise of Snark Communities

175 *The subreddit, which*: Kyler and Madison Fisher (@Fishfam), YouTube account, https://www.youtube.com/channel/UCJTyunmsBLj20wyguh6uMig.

175 *Madison Fisher*: Madison Fisher (@madisonbontempo), Instagram account, https://www.instagram.com/madisonbontempo/.

175 *Kyler Steven Fisher:* Kyler Steven Fisher (@kylerstevenfisher), Instagram account, https://www.instagram.com/kylerstevenfisher/.

175 *Taytum and Oakley Fisher*: Taytum and Oakley Fisher (@taytumandoakley), Instagram account, https://www.instagram.com/taytumandoakley/.

175 *The subreddit, which*: Halston Blake (@halston.blake), Instagram account, https://www.instagram.com/halston.blake/.

175 *Oliver and Cohen Fisher:* Oliver and Cohen Fisher (@oliverandcohen), Instagram account, https://www.instagram.com/oliverandcohen/.

176 *Four years into*: FishFamSnark (r/FisherFamilySnark), Reddit page, https://www.reddit.com/r/FisherFamilySnark/.

176 *She also frequented*: LaBrantFamSnark (@r/LaBrantFamSnark), Reddit page, https://www.reddit.com/r/LaBrantFamSnark/.

178 *r/blogsnark, which has*: Blogsnark (r/blogsnark), Reddit page, https://www.reddit.com/r/blogsnark/.

178 *There's r/doughertydozen*: doughertydozen (r/doughertydozen), Reddit page, https://www.reddit.com/r/doughertydozen/; McKnightFamSnark (r/McKnightFamSnark), Reddit page, https://www.reddit.com/r/McKnightFamSnark/; 8passengersnark (r/8passengersnark), Reddit page, https://www.reddit.com/r/8passengersnark/.

178 *On r/GDSnark, about*: GDSnark (r/GDSnark), Reddit page, https://www.reddit.com/r/GDSnark/.

180 *As the amount*: Lynn McCutcheon and Mara Aruguete, "Is Celebrity Worship Increasing Over Time?," *Journal of Studies in Social Sciences and Humanities* 7, no. 1 (April 2021): 66–75, https://www.researchgate.net/publication/351075749_Is_Celebrity_Worship_Increasing_Over_Time.

181 *"Hey, what are"*: Kaitlyn Schultz (@kaitlyn.a.schultz), TikTok account, https://www.tiktok.com/@kaitlyn.a.schultz.

181 *But maybe it's*: Jonathan Gray, "Anti-fandom and the Moral Text: Television Without Pity and Textual Dislike," *American Behavioral Scientist* 48, no. 7 (March 2005), https://doi.org/10.1177/00027642042731.

182 *Caitlin (a pseudonym)*: LaBrantFamSnark (@r/LaBrantFamSnark), Reddit page, https://www.reddit.com/r/LaBrantFamSnark/; Cole and Savannah LaBrant (@ColeAndSav), YouTube account, https://www.youtube.com/channel/UC4-CH0epzZpD_ARhxCx6LaQ.

186 *When I ask Bekah*: Bekah Martinez (@bekah), Instagram account, https://www.instagram.com/bekah/.

186 *(Bekah is mentioned)*: The Bachelor (r/thebachelor), Reddit page, https://www.reddit.com/r/thebachelor/.

187 *When I ask @annalee15*: Annalee Grace (@annalee15), Instagram page, https://www.instagram.com/annalee15/. Annalee Grace (@annaleegrace15), TikTok account, https://www.tiktok.com/@annaleegrace15.

188 *Abha Ahad, a journalist*: "in defense of snark subreddits," *Substack*, November 20, 2024, https://www.girlonline.in/p/in-defense-of-snark-subreddits.

189 *There was also*: Lunden and Olivia Stallings (@lundenandolivia), TikTok account, https://www.tiktok.com/@lundenandolivia.

189 *The weekend of*: Lundenandoliviasnark (r/Lundenandoliviasnark), Reddit page, https://www.reddit.com/r/Lundenandoliviasnark/.

189 *"Across various snark"*: Ivy Griffith, "Weight-Loss Controversy: Liv Schmidt's TikTok Sparks Mix of Fury and Praise," *Distractify*, September 19, 2024, https://www.distractify.com/p/liv-schmidt-tiktok-controversy.

190 *On r/kaitlynschultzsnarkk, one*: kaitlynschultzsnarkk (r/kaitlynschultzsnarkk), Reddit page, https://www.reddit.com/r/kaitlynschultzsnarkk/.

11. Baby's First Sponsorship:
The Economics of Influence

191 *As of this writing*: Alexandra Sabol (@alexbabii97), TikTok account, https://www.tiktok.com/@alexbabii97.

193 *According to Goldman*: Goldman Sachs Creator Economy Report, "The Creator Economy Could Approach Half-a-Trillion Dollars by 2027," Goldman Sachs, April 19, 2023, https://www.goldmansachs.com/insights/articles/the-creator-economy-could-approach-half-a-trillion-dollars-by-2027.

193 *A few years ago*: Lottie Weaver (@lottie..weaver), TikTok account, https://www.tiktok.com/@lottie..weaver.

194 *They now have*: The Bee Family (@ehbeefamily), YouTube account, https://www.youtube.com/@ehbeefamily; The Bee Family (@ehbeefamily), Instagram account, https://www.instagram.com/ehbeefamily/.

199 *Brand deals are*: Goldman Sachs Creator Economy Report, "The Creator Economy Could Approach Half-a-Trillion Dollars by 2027," Goldman Sachs, April 19, 2023, https://www.goldmansachs.com/insights/articles/the-creator-economy-could-approach-half-a-trillion-dollars-by-2027.

203 *Research backs Laskey*: Patrick van Kessel, Skye Toor, and Aaron Smith, "Children's Content, Content Featuring Children and Video Games Were Among the Most-Viewed Video Genres," Pew Research Center, July 25, 2019, https://www.pewresearch.org/internet/2019/07/25/childrens-content-content-featuring-children-and-video-games-were-among-the-most-viewed-videos-genres/.

204 *Laskey estimates that*: Maia Knight (@MaiaKnight), TikTok account, https://www.tiktok.com/@maiaknight.

204 *A source I*: Fortesa Latifi, "Parenting Influencers Try Something New: Giving Their Kids Privacy," *Washington Post*, August 1, 2023, https://www.washingtonpost.com/parenting/2023/08/01/parenting-influencers-children-privacy/.

205 *Katie Beach, a*: Katie Beach (@thekatiebeach), Instagram account, https://www.instagram.com/thekatiebeach/; Katie Beach (@thekatiebeach), TikTok account, https://www.tiktok.com/@thekatiebeach.

206 *In the 2024*: Ingrida Behri, "The Use of Children as Influencers and the Harmful Effects on their Health," *Interdisciplinary Journal of Research and Development* 11, no. 2 (July 2024), DOI:10.56345/ijrdv11n207.

206 *According to a*: Tajammul Pangarkar, "Baby Products Statistics 2025 by Wide Range," *Market.us News*, January 13, 2025, https://www.news.market.us/baby-products-statistics/.

208 *Kaci, who with*: Casey and Kaci (@caseyandkaci), TikTok account, https://www.tiktok.com/@caseyandkaci.

208 *Adrea Garza, the*: Adrea Garza (@garzacrew), TikTok account, https://www.tiktok.com/@garzacrew.

12. The Wrong Kind of Attention:
Online Predators and Invisible Threats

211 *In 2022, Wren*: Wren Eleanor (@wren.eleanor), TikTok account, https://www.tiktok.com/@wren.eleanor?lang=en.

211 *There was even*: Sarah Adams (@mom.uncharted), TikTok account, https://www.tiktok.com/@mom.uncharted.

212 *According to* Rolling: E. J. Dickson, "A Toddler on TikTok Is Spawning a Massive Mom-Led Movement," *Rolling Stone*, July 20, 2022, https://www.rollingstone.com/culture/culture-news/tiktok-wren-eleanor-moms-controversy-1385182/.

212 *When I reached*: TikTok Community Guidelines, "Youth Safety and Well-Being," TikTok (web page), https://www.tiktok.com/community-guidelines/en/youth-safety?lang=en.

212 *"We do not"*: TikTok Community Guidelines, "Sensitive and Mature Themes," TikTok (web page), https://www.tiktok.com/community-guidelines/en/sensitive-mature-themes.

212 *Meta's community guidelines*: "Online child protection," *Meta*, https://about.meta.com/actions/safety/onlinechildprotection/.

214 *In February of*: Jennifer Valentino-DeVries and Michael H. Keller, "A Marketplace of Girl Influencers Managed by Moms and Stalked by Men," *New York Times*, February 22, 2024, https://www.nytimes.com/2024/02/22/us/instagram-child-influencers.html.

215 *More than a risk*: Elisabeth Van den Abeele et al., "Managing Authenticity in a Kidfluencers' World: A Qualitative Study with Kidfluencers and Their Parents," *New Media & Society* 27, vol. 6 (January 2024): 3240–63, https://doi.org/10.1177/14614448231222558.

215 *Additionally, researchers have*: Elisabeth Van den Abeele et al., "Child's Privacy Versus Mother's Fame: Unravelling the Biased Decision-Making Process of Momfluencers to Portray Their Children Online," *Information, Communication & Society* 27, no. 2 (October 2022): 297–313, https://doi.org/10.1080/1369118X.2023.2205484.

215 *And the problem*: Cecilia Kang, "A.I.-Generated Images of Child Sexual Abuse Are Flooding the Internet," *New York Times*, July 10, 2025, https://www.nytimes.com/2025/07/10/technology/ai-csam-child-sexual-abuse.html.

216 *Adrea Garza, the*: Adrea Garza (@garzacrew), TikTok account, https://www.tiktok.com/@garzacrew.

219 *In the Kids*: Francis Rees, "'Kids as Content' Digital Safeguarding Toolkit," University of Essex, May 19, 2025, https://www.essex.ac.uk/research-projects/kids-as-content.

219 *When I spoke to Lottie*: Lottie Weaver (@lottie..weaver), TikTok account, https://www.tiktok.com/@lottie..weaver.

220 *In the study*: Emma Beuckels et al., "Work It Baby! A Survey Study to Investigate the Role of Underaged Children and Privacy Management Strategies Within Parent Influencer Content," *New Media & Society* 27, no. 6 (January 2024): 3081–101, https://doi.org/10.1177/14614448231218992.

220 *Another study found*: Elisabeth Van den Abeele et al., "Child's Privacy Versus Mother's Fame: Unravelling the Biased Decision-Making Process of Momfluencers to Portray Their Children Online," *Information, Communication & Society* 27, no. 2 (October 2022): 297–313, https://doi.org/10.1080/1369118X.2023.2205484.

220 *Jess Spaulding*: Jess Spaulding (@jess.spaulding), TikTok account, https://www.tiktok.com/@jess.spaulding.

13. Lights, Camera, Legislation:
Regulating Family Vlogging

225 *In 2023, Illinois*: Fortesa Latifi, "Illinois Just Passed the Country's First Law Protecting Children of Influencers," *Teen Vogue*, August 11, 2023, https://www.teenvogue.com/story/illinois-child-influencers-law.

225 *According to the Associated Press*: Claire Savage, "Starting Next Year, Child Influencers Can Sue If Earnings Aren't Set Aside, Says New Illinois Law," Associated Press, August 12, 2023, https://apnews.com/article/tiktok-child-influencer-illinois-social-media-f784b4bc52cb75ad1e0d28785993b1c5.

226 *And the law*: April Rubin, "Classic Hollywood Laws Inspire Shields for Kids in Multibillion Influencer Industry," *Axios*, July 27, 2024, https://www.axios.com/2024/07/27/child-influencer-protections-illinois-minnesota.

226 *Similar to the*: HF 3488 3rd Engrossment, 93rd Legislature (2023–2024), https://www.revisor.mn.gov/bills/text.php?number=HF3488&version=latest&session=ls93&session_year=2024&session_number=0.

226 *I talked to*: Fortesa Latifi, "Influencer Parents and the Kids Who Had Their Childhood Made into Content," *Teen Vogue*, March 10, 2023, https://www.teenvogue.com/story/influencer-parents-children-social-media-impact.

227 *In September of*: California Senate Bill No. 764, "Minors: Online Platforms," Chapter 611, September 27, 2024, https://leginfo.legislature.ca.gov/faces/billTextClient.xhtml?bill_id=202320240SB764; Fortesa Latifi, "California Passes Legal and Financial Protections for Child Influencers,"

Teen Vogue, September 26, 2024, https://www.teenvogue.com/story/california-legislation-for-kid-influencers.

228 *I spoke to*: Coogan Law, SAG-AFTRA (website), https://www.sagaftra.org/membership-benefits/young-performers/coogan-law.

228 *In October of 2024*: Fortesa Latifi, "Utah Joins the Debate Over Child Influencer Rights," *Washington Post*, October 16, 2024, https://www.washingtonpost.com/lifestyle/2024/10/16/utah-protections-children-mom-influencers/.

229 *In her statement*: Shari Franke, "Attention Utah Residents: I Have Been Working on Drafting HB322 That Would Protect Child Influencers in Our State," Shari Franke, Instagram post, February 12, 2025, https://www.instagram.com/p/DF_g24xssZW/.

231 *In March of*: Utah State Legislature, HB 322, Child Actor Regulations, March 25, 2025, https://le.utah.gov/~2025/bills/static/HB0322.html.

231 *In my reporting*: Fortesa Latifi, "Utah Passed a Law to Protect Child Influencers," *Teen Vogue*, March 25, 2025, https://www.teenvogue.com/story/utah-passed-a-law-to-protect-child-influencers.

232 *In July of*: Virginia State Legislature, HB 2401, May 2, 2025, https://lis.virginia.gov/bill-details/20251/HB2401/text/CHAP0699.

233 *"No one really"*: Francis Rees, "Child Influencer Project—Ireland and the UK," University of Essex, March 6, 2024, https://www.essex.ac.uk/research-projects/child-influencer-project.

233 *That can be*: Annie Dunn, "The Kids Aren't Alright: Creating Greater Protections for the Children of Family Vloggers," *Oklahoma Law Review* 77, no. 2 (February 2025), https://digitalcommons.law.ou.edu/olr/vol77/iss2/8.

234 *Let's take a*: Utah State Legislature, HB 322, Child Actor Regulations, March 25, 2025, https://le.utah.gov/~2025/bills/static/HB0322.html.

237 *"Under the Utah"*: Taylor Lorenz, "New Law 'Protecting Child Influencers' Is Bad and Dangerous," *User Mag* (blog), April 14, 2025, https://www.usermag.co/p/utah-protect-child-influencers-law-bad-and-dangerous-free-speech-censorship-concerns-legal-experts.

238 *In a research*: Francis Rees, "Famous at Five: Risk Assessing Digital Child Labour," *Information & Communications Technology Law* (January 2025): 1–22, https://doi.org/10.1080/13600834.2025.2452699.

Index

About the Author

Fortesa Latifi is a journalist who has written for *Rolling Stone*, *The Cut*, *The New York Times*, and more. She lives in Los Angeles with her husband and daughter. You can find her online @hifortesa.